183
Advances in Polymer Science

AF789351

Advances in Polymer Science

Recently Published and Forthcoming Volumes

Crosslinking in Materials Science
Vol. 184, 2005

Phase Behavior of Polymer Blends
Volume Editor: Freed, K.
Vol. 183, 2005

Polymer Analysis/Polymer Theory
Vol. 182, 2005

Interphases and Mesophases in Polymer Crystallization II
Volume Editor: Allegra, G.
Vol. 181, 2005

Interphases and Mesophases in Polymer Crystallization I
Volume Editor: Allegra, G.
Vol. 180, 2005

Inorganic Polymeric Nanocomposites and Membranes
Vol. 179, 2005

Polymeric and Inorganic Fibres
Vol. 178, 2005

Poly(arylene Ethynylenes)
From Synthesis to Application
Volume Editor: Weder C.
Vol. 177, 2005

Ring Opening Metathesis
Volume Editor: Buchmeister, M.
Vol. 176, 2005

Polymer Particles
Volume Editor: Okubo M.
Vol. 175, 2005

Neutron Spin Echo in Polymer Systems
Authors: Richter, D., Monkenbusch, M., Arbe, A., Colmenero, J.
Vol. 174, 2005

Advanced Computer Simulation Approaches for Soft Matter Sciences I
Volume Editors: Holm, C. Kremer, K.
Vol. 173, 2005

Microlithography · Molecular Imprinting
Vol. 172, 2005

Polymer Synthesis
Vol. 171, 2004

NMR · Coordination Polimerization · Photopolymerization
Vol. 170, 2004

Long-Term Properties of Polyolefins
Volume Editor: Albertsson A.-C.
Vol. 169, 2004

Polymers and Light
Volume Editor: Lippert T. K.
Vol 168, 2004

New Synthetic Methods
Vol. 167, 2004

Polyelectrolytes with Defined Molecular Architecture II
Volume Editor: Schmidt M.
Vol. 166, 2004

Polyelectrolytes with Defined Molecular Architecture I
Volume Editor: Schmidt M.
Vol. 165, 2004

Filler-Reinforced Elastomers · Scanning Force Microscopy
Vol. 164, 2003

Liquid Chromatography · FTIR Microspectroscopy · Microwave Assisted Synthesis
Vol. 163, 2003

Phase Behavior of Polymer Blends

Volume Editor: Karl Freed

With contributions by
N. Clarke · J. Dudowicz · K. F. Freed · D. Schwahn

Springer

This series presents critical reviews of the present and future trends in polymer and biopolymer science including chemistry, physical chemistry, physics and material science. It is adressed to all scientists at universities and in industry who wish to keep abreast of advances in the topics covered.

As a rule, contributions are specially commissioned. The editors and publishers will, however, always be pleased to receive suggestions and supplementary information. Papers are accepted for "Advances in Polymer Science" in English.

In references Advances in Polymer Science is abbeviated *Adv Polym Sci* and is cited as a journal.

The electronic content of *Adv Polym Sci* may be found at springerlink.com

ISSN 0065-3195
ISBN 3-642-42420-1 **Springer Berlin Heidelberg New York**
ISBN 978-3-642-42420-5 **Springer Berlin Heidelberg New York**
DOI 10.1007/b135844

Springer is a part of Springer Science+Business Media
springeronline.com

Softcover re-print of the Hardcover 1st edition 2005

Cover design: *Design & Production* GmbH, Heidelberg
Typesetting and Production: LE-TEX Jelonek, Schmidt & Vöckler GbR, Leipzig

Printed on acid-free paper 02/3141 YL – 5 4 3 2 1 0

Volume Editors

Prof. Karl F. Freed

James Franck Institute and Department of Chemistry
The University of Chicago
Chicago IL, 60637, USA
k-freed@uchicago.edu

Prof. Akihiro Abe

Department of Industrial Chemistry
Tokyo Institute of Polytechnics
1583 Iiyama, Atsugi-shi 243-02, Japan
aabe@chem.t-kougei.ac.jp

Prof. A.-C. Albertsson

Department of Polymer Technology
The Royal Institute of Technology
S-10044 Stockholm, Sweden
aila@polymer.kth.se

Prof. Ruth Duncan

Welsh School of Pharmacy
Cardiff University
Redwood Building
King Edward VII Avenue
Cardiff CF 10 3XF
United Kingdom
duncan@cf.ac.uk

Prof. Karel Dušek

Institute of Macromolecular Chemistry, Czech
Academy of Sciences of the Czech Republic
Heyrovský Sq. 2
16206 Prague 6, Czech Republic
dusek@imc.cas.cz

Prof. Dr. W. H. de Jeu

FOM-Institute AMOLF
Kruislaan 407
1098 SJ Amsterdam, The Netherlands
dejeu@amolf.nl
and
Dutch Polymer Institute
Eindhoven University of Technology
PO Box 513
5600 MB Eindhoven, The Netherlands

Prof. Jean-François Joanny

Physicochimie Curie
Institut Curie section recherche
26 rue d'Ulm
F-75248 Paris cedex 05, France
jean-francois.joanny@curie.fr

Prof. Hans-Henning Kausch

EPFL SB ISIC GGEC
J2 492 Bâtiment CH
Station 6
CH-1015 Lausanne, Switzerland
kausch.cully@bluewin.ch

Prof. S. Kobayashi

Department of Materials Chemistry
Graduate School of Engineering
Kyoto University
Kyoto 615-8510, Japan
kobayasi@mat.polym.kyoto-u.ac.jp

Advances in Polymer Science
Also Available Electronically

For all customers who have a standing order to Advances in Polymer Science, we offer the electronic version via SpringerLink free of charge. Please contact your librarian who can receive a password or free access to the full articles by registering at:

springerlink.com

If you do not have a subscription, you can still view the tables of contents of the volumes and the abstract of each article by going to the SpringerLink Homepage, clicking on "Browse by Online Libraries", then "Chemical Sciences", and finally choose Advances in Polymer Science.

You will find information about the

- Editorial Board
- Aims and Scope
- Instructions for Authors
- Sample Contribution

at springeronline.com using the search function.

Preface

The observation that blending existing polymers can produce new and improved materials with the beneficial features of each component has focused attention on understanding the phase behavior of polymer blends because the resulting composites may form homogeneous phases or may be phase separated with a particular desirable morphology. While Flory–Huggins theory correctly predicts that liquid-state polymers generally do not tend to mix, its predictive ability for designing these new materials are rather limited. Thus, in order to rectify this deficiency in the theory, several approaches have been used to develop more ambitious theories, while experimental efforts have been devoted to devising various strategies for improving mixing including, for instance, the use of flow, as described in the article by Clarke in this volume. Thus, progress in elucidating the factors affecting and promoting blend miscibility has benefited from strong interactions between experimental and theoretical groups, as emphasized by the articles herein.

The subject of the phase behavior of polymer blends is quite vast and much too large to fit in a single volume of Advances in Polymer Science, so the present volume is the first in a planned series. The lead article in this volume by Schwahn describes his small angle neutron scattering experiments for a variety of polymer blend systems, including homopolymer and copolymer blends, as well as the fascinating mixtures of diblock copolymers with homopolymers and the technologically important influences of pressure and additives on blend phase behavior. Schwahn's article focuses on the important influences of critical fluctuations on the phase behavior and emphasizes the need for analyzing experimental data accordingly in order to provide meaningful comparisons between experiments and theories of polymer blends, which are generally of the mean field variety. The second article by Freed and Dudowicz is devoted to a description of several applications of the lattice cluster theory (LCT) in the simplified high molecular weight, incompressible limit to demonstrate how monomer molecular structure strongly affects the phase behavior of polymer blends. Examples of LCT predictions that can not be obtained from Flory–Huggins theory involve, for instance, the possibility of lower critical solution temperature phase behavior for incompressible binary blends and microphase ordering of diblock copolymers upon heating. When applied to copolymer systems, LCT provides a significant extension of Flory–Huggins type theories for copolymers by incorporating the important contributions from the temperature-independent portion of the Flory effective interaction parameter. Several direct comparisons of the LCT to experiments serves to illustrate the utility of this simple approach to interpret measurements and to elucidate the physics governing blend miscibility. The last article by Clarke describes both theoretical methods and experimental observations of the influence of shear flow on the properties of polymer blends. Comparisons are made between the predictions of two conflicting theories: a quasiequilibrium theory in which a shear-dependent contribution is appended to the free energy, and a theory based on the modification of the equations of motion by the presence of the shear flow. Clarke discusses several of the challenges towards further theoretical progress.

Chicago, June 2005 Karl. F. Freed

Contents

Adv Polym Sci (2005) 183: 1–61
DOI 10.1007/b135882

Published online: 26 July 2005

Critical to Mean Field Crossover in Polymer Blends

Dietmar Schwahn

Institut für Festkörperforschung, Forschungszentrum Jülich GmbH, 52425 Jülich, Germany
d.schwahn@fz-juelich.de

Abstract Crossover phenomena due to thermal composition fluctuations play a multifarious role in polymer blends. This is demonstrated in this article by describing results from small angle scattering experiments in particular with neutrons. Scattering methods are a direct tool to measure the strength and correlation length of composition fluctuations. We will review the effects of thermal fluctuations in binary (A/B) polymer blends under various conditions of external temperatures and pressures and additives, such as non-selective solvents and (A-B) diblock copolymers, and will give an interpretation with the corresponding crossover theories. General conclusions are that the effects from thermal composition fluctuations have to be more seriously considered in polymer blends and that the more sophisticated crossover theories are needed for a precise determination of the Flory–Huggins interaction parameter and the phase boundaries. In addition, we discuss observations of crossover to other universality classes such as the 3D-Ising case, namely the transition to the renormalized Ising case when the composition of a third component starts to fluctuate and to the isotropic Lifshitz critical behavior when an (A-B) diblock copolymer is added.

Keywords Polymer blends · Crossover by thermal composition fluctuations · 3D-Ising critical behavior · Renormalized Ising critical behavior · Lifshitz critical behavior · Ginzburg criterion · Flory–Huggins parameter · External fields of temperature and pressure · Additives of non-selective solvent and diblock copolymers

1 Introduction

Polymer blends represent materials for a large class of industrial products and are the subject of intensive academic research. Blending of chemically different polymers is an important tool in industrial production for tailoring products with optimized material properties [1]. On the other hand polymer mixtures are model systems in statistical physics for studying fundamental aspects of equilibrium and non-equilibrium properties such as phase diagrams, thermal composition fluctuations far and near the critical point of a second order phase transition, conformational properties of the chains, the kinetics of phase transitions as well as the detailed dynamics of diffusion processes. There are strong activities in this field from both theory and experiment as can be realized from numerous review articles [2–5].

In this article we will focus on the behavior of thermal composition fluctuations in binary (A/B) homopolymer blends (A, B denote the monomers) in different external fields such as temperature and pressure, in their pure state, as well as with small additions of a non-selective solvent and symmetric (A-B) diblock copolymers. The external field conditions as well as additives can strongly influence the degree of thermal fluctuations and lead to crossover phenomena between different universality classes of mean field, 3D-Ising, renormalized 3D-Ising, and isotropic Lifshitz critical behavior. Each universality class is characterized by a unique set of critical exponents describing the divergence of several thermodynamic parameters in the near vicinity of a second order phase transition. Two of those prominent parameters, namely, the correlation length of thermal composition fluctuations and the corresponding susceptibility are determined in scattering experiments. Generally, scattering methods are a sensitive tool to measure spatial inhomogeneities, such as thermal composition fluctuations in blends, and neutrons have the particularly important advantages of strong scattering contrast and deep penetration into materials [6, 7]. So, it is quite natural that most of the relevant studies in this field have been performed with neutron small angle scattering (SANS) techniques.

Thermal composition fluctuations are always present in multi-component systems. If the interaction energy between same components is stronger, clustering of (A/A) and (B/B) monomers is preferred, and phase separation will occur at low temperatures. A schematic phase diagram of a symmetric binary polymer blend is depicted in Fig. 1. It represents an "upper critical solution temperature" (UCST) system; at high temperatures the blend is homogeneously mixed, while at low temperatures the system decomposes into macroscopically large domains being rich in A or B polymers. Such a pro-

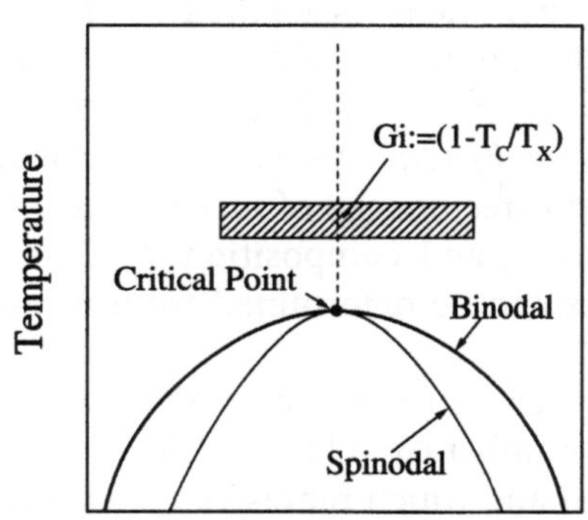

Fig. 1 Schematic phase diagram of binary polymer melt of equal molar volume $V_A = V_B$. At high temperature both polymers are miscible; at low temperature below the binodal the sample separates into two macroscopic large domains with compositions determined by the binodal. The spinodal separates metastable and unstable regions which determine the process of phase separation. The disordered regime is separated by the Ginzburg number *Gi* domain into regions of small and large degrees of thermal composition fluctuations

cess can be easily followed by eye as the sample becomes turbid from the strong scattering of light by the μm large domains. The binodal is the border line between the one and two-phase regimes, while the spinodal separates the meta-stable and unstable parts of the two-phase regime. The binodal and spinodal meet at the critical point which also represents the highest temperature of immiscibility in monodisperse polymer systems [8]. In symmetric blends with equal molar volumes $V_A = V_B$ theory predicts a critical composition of $\Phi_C = 50\%$ volume fraction. The one-phase regime is homogeneous on a macroscopic length scale and usually appears transparent to the eye. On smaller length scales, however, the composition is heterogeneous due to thermal fluctuations. The occurrence of thermal fluctuations is a dynamic process as they are created with a given probability and decay afterwards with a relaxation rate determined by the interdiffusion constant [3, 4]. In static elastic neutron scattering experiments one measures the equilibrium mean square deviation $\langle \delta\Phi^2 \rangle$ from the average composition Φ given by the volume fraction of one of the components. The range of those fluctuations is determined by the correlation length ξ which sensitively depends on external parameters such as temperature and pressure: ξ is of the order of interatomic distances at high temperatures when entropy is dominating; it is large at low temperatures and becomes infinite at the critical point. The critical point represents a particular position of the phase boundary where a continuous (second order) phase transition is observed.

According to the fluctuation–dissipation theorem the equilibrium mean square deviation of thermal composition fluctuations $\langle \delta\Phi^2 \rangle$ is related to the first derivative of the order parameter with respect to the chemical potential [3, 4]. If the order parameter is defined as the composition Φ of the component "A" then the conjugate field is represented by the difference of the chemical potentials $\Delta\mu(= \mu_A - \mu_B)$; so the degree of thermal fluctuations is related to $\partial_{\Delta\mu}\Phi(\partial_{\Delta\mu} \equiv \partial/\partial\Delta\mu)$ which is equivalent to $1/\partial^2_\Phi \Delta G$, where ΔG represents the Gibbs free energy of mixing with its natural parameters temperature T, pressure P, and composition Φ. This shows how the thermodynamic parameters can be determined from measurements of thermal fluctuations.

The border line between strong and weak thermal fluctuations is estimated by the Ginzburg number $Gi := (1 - T_C/T_X)$, which is evaluated from the Ginzburg criterion [3, 4] and which represents a reduced temperature T_X below which deviations from the mean field approach are observed (see hatched area in Fig. 1). Fluctuations are considered to be weak as long as the fluctuation modes superimpose linearly. In this weak limit the fluctuation modes can be described within the Gaussian approximation, and mean field theory is a good approximation. Near the critical point thermal fluctuations become strong and lead to visible non-linear effects. In this range more sophisticated theories as the 3D-Ising model and crossover theories are needed [9, 10].

Fluctuation effects are usually neglected in theoretical descriptions of polymer blends which are therefore described within the mean field theoretical approach of the Flory–Huggins (FH) theory [2–4]. In this model the Gibbs free energy of mixing ΔG is depicted by a combinatorial entropy of mixing being inversely proportional to the molar volume V and by the FH parameter $\Gamma = \Gamma_h/T - \Gamma_\sigma$ being described by enthalpic and non-combinatorial entropic terms Γ_h and Γ_σ, respectively. The general neglect of thermal fluctuations in polymer blends might be explained with the original estimate of the Ginzburg criterion by deGennes on the basis of incompressible FH theory. This relationship represents a universal criterion according to $Gi \propto 1/N$ (N degree of polymerization) and proposes an extremely small Ginzburg number [3]. Since in low molecular liquids a $Gi \cong 10^{-2}$ is the expected value, a roughly N times smaller Gi is estimated in polymer blends indicating a small region with strong thermal fluctuations [10]. Meanwhile, it has been shown that the Ginzburg criterion for polymer blends is non-universal with appreciably larger Ginzburg numbers and therefore larger regions where thermal fluctuations are significant. Theoretical considerations, computer simulations, and scattering experiments show that the compressibility is dependent on the non-combinatorial entropy Γ_σ which has a strong influence on the Ginzburg criterion and thereby on the degree of thermal composition fluctuations. The effect of thermal fluctuations will be the main emphasis of this article.

The influence of several chain parameters on Γ_σ and therefore on Gi has been identified. These are polymer asymmetry, different monomer structures, chain stiffness, compressibility, and chain end-effects. The effects of compressibility and monomer structure have been discussed by Lifshitz et al. [11] and Dudowicz et al. [12] on basis of the Lattice Cluster Theory (LCT), respectively. Considering these effects a one to several orders of magnitude increase of Gi is found. The present theoretical ideas behind an extended Ginzburg criterion have in common the consideration of correlations (or non-random mixing) on length scales between the coil and monomer size. These shorter length scale fluctuations lead to an extended mean field description and are comprised into a non-combinatorial entropic term Γ_σ of the FH parameter. It will be shown later (Eq. 18) in the context of the crossover theories for the susceptibility and correlation length that Γ_σ strongly influences the Ginzburg number and thereby the strength of thermal fluctuations and the domain over which they must be considered in the analysis of experimental data.

Another theoretical approach including the effects of thermal fluctuations has been derived from field-theoretical methods (see recent review by Fredrickson et al. [13]). Wang [14] quite recently gave a compact overview and a derivation from a systematic renormalization procedure of an effective FH parameter and Ginzburg criterion. Starting from a "bare" FH parameter as originally derived by Flory, which only considers the microscopic enthalpic interaction, Wang formulated an "effective" FH parameter comprising the effects of local correlations of wavelengths between monomer and chain sizes. The last mentioned parameter is determined from SANS experiments within a mean field approximation, is comparable with the FH parameter derived from the LC-theory, and also includes molecular conformational asymmetry effects as discussed in [15]. In addition, an "apparent" FH parameter is determined which also includes contributions from long wavelength fluctuations that become relevant near the critical point. Fluctuations lead to a reduced effective FH parameter, to a decrease of the critical temperature, and to a larger *Gi*. In the limit of large degree of polymerization, a $1/N$ scaling of *Gi* is predicted.

Monte Carlo simulations very early demonstrated the effect of thermal composition fluctuations in low molecular blends. Studies by Sariban et al. [16] exclusively found Ising critical behavior in blends of molar volume up to about 16 000 cm^3/mol and no indications of a crossover to mean field behavior. Such a mean field crossover was later detected by Deutsch et al. [17] in blends with an order of magnitude larger chains. These results and the techniques of Monte Carlo simulations have been extensively reviewed by Binder in [4].

These theoretical considerations and Monte Carlo simulations were accompanied by scattering experiments, which consistently show that the range of relevant thermal fluctuations and thereby *Gi* is at least of the order of magnitude larger than originally estimated from simple incompressible FH theory [18–26]. An important development in the description of thermal fluctuations is the derivation of proper analytic crossover functions for the susceptibility and correlation length which also can be relatively easily handled by the experimentalist [10, 27–30]. These equations allow an analysis of the susceptibility and correlation length over the whole experimental range including mean field and Ising behavior and therefore lead to much more precisely determined thermodynamic parameters. A Ginzburg criterion is quite naturally derived from these crossover functions and is determined by the ratio of the mean field and Ising critical amplitudes for the susceptibility and/or correlation length. The so determined Ginzburg number is a non-universal function, which sensitively depends on the degree of polymerization and on the FH parameter non-combinatorial entropy of mixing Γ_σ [31, 32]. An increase of Γ_σ leads to a strong increase of *Gi*, and only in the limit of negligible Γ_σ is a universal Ginzburg criterion with the proposed $1/N$ scaling behavior obtained consistent with earlier predictions for incompressible polymer blends.

High external pressure fields can lead to an appreciable change of the FH interaction parameter, the phase boundaries, and the Ginzburg number. Pressure usually induces an increase of the phase transition temperature, a decrease of the Ginzburg parameter and of the enthalpic (Γ_h) and absolute value of the entropic (Γ_σ) terms of the FH parameter. An increase of the phase boundary with pressure is expected as it is related to a decrease of the free volume and a corresponding decrease of the entropic Γ_σ [33–41]. In LCST systems (phase transition occurs at high temperatures) such as the PS/PVME (polystyrene/poly(vinyl methyl ether) blend Γ_h is constant within 1 and 120 MPa while Γ_σ is negative and increases with pressure [34]. In a few blends, however, an "abnormal" pressure induced decrease of the phase boundary and a corresponding increase of the Γ_h and Γ_σ FH terms are observed [42, 43]. The shift of the phase boundary with pressure is described by the Clausius–Clapeyron equation in terms of the FH parameter and the Ginzburg number [40]. In some blends one actually finds a shift of the critical temperature dominated by the Ginzburg number according to $\partial_P Gi$ [40]. The exploitation of the Clausius–Clapeyron equation also allows a check of consistency with respect to the underlying theory, such as the dependence of the Ginzburg criterion on the entropic FH term Γ_σ. In this respect, the study of pressure induced changes of thermal fluctuations may lead to relevant insight into the mechanisms governing the polymer blend thermodynamic properties and their interrelations.

Mixing a binary polymer blend with a small amount of a third component usually leads to strong changes of the phase behavior and in some cases even to a crossover to a different universality class. The addition of a non-selective solvent generally leads to improved compatibilization and to a larger Ginzburg number. Furthermore, very near the critical temperature, a crossover from Ising to the renormalized Ising behavior is predicted for those blend-solvent systems. Such crossover phenomena have been intensively explored by the group of Nose [44, 45] with light scattering. They found this type of crossover very pronounced after adding a selective solvent but only rather weak in a few special cases after adding a non-selective solvent. These observations are consistent with the SANS studies discussed in this article.

A much more complex situation is achieved if an (A-B) diblock copolymer is mixed with a binary (A/B) homopolymer blend. Generally the diblock copolymer leads to a better compatibilization of the homopolymers similar to the action of surfactant molecules in oil-water mixtures. A complex phase diagram is obtained that depends upon the diblock content: Several disordered and ordered phases appear at, respectively, high and low temperatures. At low and high diblock content a two-phase region of macroscopically large domains and an ordered phase known from diblock copolymer melts emerge, respectively, and intermediate diblock content leads to droplet and bicontinuous microemulsion phases [46–52]. Bicontinuous microemulsion and ordered phases at high diblock copolymer concentrations are characterized by a periodicity length Λ which in scattering experiments becomes visible as a peak at $Q^* = 2\pi/\Lambda$. Within the disordered phases several crossover phenomena are observed, namely, from mean field to 3D-Ising and to isotropic Lifshitz critical behavior, as well as to the Brasovskii type of pure diblock copolymers [53]. One reason for this complex behavior is that homopolymer mixtures and diblock copolymers belong to different universality classes. Diblock copolymers shows much stronger thermal fluctuations which even change the character of the disorder-order phase transition from second to first order and lead to a broader critical range with a weaker N proportionality of Gi according to $Gi \propto 1/\sqrt{N}$ [54].

Mean field theory predicts that the critical lines of "blend like" and "diblock like" behavior meet at the isotropic critical Lifshitz point and the Lifshitz line (LL) which is defined when Q^* becomes zero. The isotropic critical Lifshitz point represents a new universality class [54–56]. Under special conditions even a tricritical Lifshitz point is predicted [55]. In this article we will discuss in some detail SANS experiments on a mixture of a critical binary (A/B) polymer blend with different concentrations of a symmetric (A-B) diblock copolymer of roughly five times larger molar volume. Under such conditions an isotropic critical Lifshitz point is predicted [55].

Near the critical Lifshitz point the phase behavior and the corresponding phase diagram are strongly influenced by thermal composition fluctuations. The strength of these fluctuations can be understood from a decrease of the surface energy acting as a restoring force for thermal fluctuations. A further peculiarity is, that a critical Lifshitz point can only exist within mean field approximation. It is "destroyed" by thermal fluctuations because the Ginzburg criterion and therefore the stabilization of the miscible phase is of different strengths on both sides of the Lifshitz line. A further observation is a change of the LL concentration with temperature near the two-phase regime [48, 49]; recent renormalization group calculations explain these observations as due to thermal fluctuations [57]. In the range near the LL one also observes a microemulsion phase not predicted by the corresponding existing mean field theories [47].

From these introductory remarks one already gets the impression about the prominent and diversified role of thermal composition fluctuations on the properties of polymer blends, which will be detailed in the following sections mainly from an experimental point of view.

2 Small Angle Neutron Scattering

A separate section is devoted to describing small angle neutron scattering (SANS) techniques as this experimental method plays a prominent role in the investigation of polymer blends. Neutrons are a particularly appropriate probe for explorations of polymer properties because of the good scattering contrast conditions and their weak interactions with material in general [6, 7]. The last mentioned condition relies on the fact that neutrons have no charge and thus allow the exploration of several mm thick samples, making SANS a non-destructive tool. The above mentioned scattering contrast is proportional to the difference of the coherent scattering length density of the polymer components and is mainly determined from the neutron interaction with the atomic nuclei. This interaction can lead to appreciably different scattering lengths for isotopes as most prominently found for hydrogen and deuterium. Consequently, a mixture of deuterated and protonated polymer chains shows a strong scattering contrast and thereby makes neutrons a sensitive tool. Beyond that the contrast variation with H/D content of selected polymers offers detailed insight into single chain properties in complex systems as the above mentioned (A/B/A-B) polymer blend-diblock copolymer mixtures.

2.1 Experimental Design of a SANS Diffractometer

A schematic lay-out and photography of the SANS experiments at the research reactor FRJ2 in Jülich are given in Fig. 2a and b. Neutrons are released in nuclear reactors by fission reactions of the uranium isotope 235 or are evaporated in spallation sources from a heavy metal target after being bombarded with high energy protons. For example about 20 neutrons are released from a lead nucleus per 1 GeV proton [7]. After being released neutrons are moderated to lower kinetic energies. For SANS and other neutron scattering methods as spin-echo and back scattering spectrometers a special "cold source" moderator of liquid hydrogen or deuterium is installed near the reactor core which delivers more than one order of magnitude larger intensity of long wavelength neutrons between 7 and 15 Å. Neutrons of this wavelength approach a smaller momentum transfer Q, as will be defined in Eq. 1, and avoid double Bragg scattering in crystalline materials. These "cold" neutrons are guided to the instrument through neutron guides that operate by total reflection. Neutron guides consist of evacuated glass channels with good surface quality and are coated with an element having a large coherent scattering length such as the Nickel 58 isotope.

The neutrons entering the instrument first pass a velocity selector, then a collimator with two apertures, a sample, and are finally detected in a position sensitive detector if scattered into a given angular interval. The selector and collimator, respectively, determine the wavelength with a relative distribution of typically $\Delta\lambda/\lambda \cong 0.10$ and the divergence of the neutron beam. In order to keep multiple scattering low, the sample should scatter only part (ca. 10%) of the incident neutrons (intensity I_0) whereas the non-scattered neutrons (intensity I_0^*) are absorbed in a beam stopper in front of the detector. The distances of sample to detector as well as of the two apertures can be changed typically from 1 to 20 m allowing the measurement of scattering angles from roughly 0.1° to 20° and to the adjustment of an optimized relationship between maximum intensity and sufficiently good resolution. In

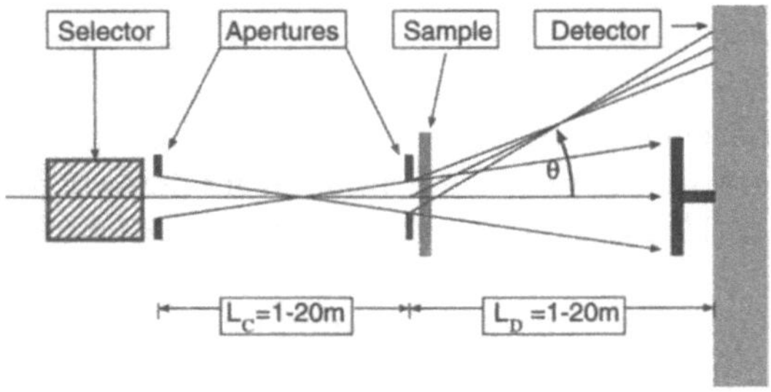

Fig. 2 *Top*: Schematic lay-out of a pin hole small angle neutron diffractometer (SANS). *Bottom*: Photography of one of the two SANS instruments at the research reactor FRJ-2 at Jülich. Vacuum sample chamber and detector tank are visible. Both Jülich SANS instruments have been in operation since 1987 and have up until now been seen worldwide to the most powerful instruments after the two SANS instruments at the ILL in Grenoble (France)

order to gain optimized intensity conditions, additional neutron guides can be inserted into the neutron beam axis so that neutrons are always transported through neutron guides until the first aperture. The scattered intensity is usually given as a function of the momentum transfer $\underline{Q}$ whose absolute value Q is determined according to

$$Q = (4\pi/\lambda)\sin(\Theta/2) \qquad (1)$$

Q is inversely proportional to the neutron wave length λ and is proportional to the scattering angle Θ. The typical wavelengths between 5 and 10 Å lead to a Q range of $10^{-3} < Q[\text{Å}^{-1}] < 0.3$ and correspond in real space to a resolution range between $10 < R[\text{Å}] < 10^3$. The Q range can be extended to smaller Q by other SANS techniques using a focusing mirror optic or perfect Silicon single crystals as monochromator and analysor [58].

The neutron intensity $I_D(\underline{Q})$ scattered by the momentum transfer $\underline{Q}$ is described in Eq. 2; it is proportional to the incident intensity I_0, the sample thickness D, the sample transmission $T = I_0^*/I_0$, and the space angle $\Delta\Omega_D$ of a detector element,

$$I_D(\underline{Q}) = I_0 DT \frac{d\Sigma}{d\Omega}(\underline{Q})\Delta\Omega_D \tag{2}$$

The transmission describes the attenuation of the neutrons non-scattered by the sample, and the macroscopic cross section $d\Sigma/d\Omega$ in units of cm^{-1} is the quantity to be determined. The calibration of the macroscopic cross section $d\Sigma/d\Omega$ in absolute units is usually performed by an additional measurement with a sample of known $d\Sigma/d\Omega$ or by measuring the direct beam in both cases with the same collimator setting [59, 60].

2.2 Scattering Cross Section of an Ideal Polymer Mixture

Blending two polymers with zero energy of mixing (FH parameter $\chi = 0$) leads to an ideal mixture. Such ideal mixtures do not show any phase behavior and can be realized within good approximation by blending protonated and deuterated polymers of the same species of not too large molar volumes V. In order to determine the structure factor $S(Q)$ of such a melt, we consider the chains on a so-called "Flory" lattice. All lattice points are of the same size and are occupied by the monomers of the two A/B polymers. The basic equation for the macroscopic cross section in SANS within first Born approximation is given according to [6, 7]

$$d\Sigma/d\Omega(\underline{Q}) = (1/V_S)\left|\sum \bar{b}_i \exp(i\underline{Q}\underline{r}_i)\right|^2 \tag{3}$$

The scattering amplitude is determined from the sum of the coherent scattering length, $\bar{b}_i$ of the monomer at position i over all lattice points multiplied with the corresponding phase factor $\exp(i\underline{Q}\underline{r}_i)$. The square of the scattering amplitude divided by the sample volume V_S gives $d\Sigma/d\Omega(\underline{Q})$. From now on the momentum transfer will be regarded as a scalar as we always will discuss isotropically scattering samples.

To proceed, we define a variable σ_i with the following meaning,

$$\sigma_i = \begin{cases} 1 & \bar{b}_i = \bar{b}_A \\ 0 & \bar{b}_i = \bar{b}_B \end{cases} \tag{4}$$

which determines the occupation of the Flory lattice by the monomers of the polymers A and B as represented by their coherent scattering length. The scattering length at position i is then given as $\bar{b}_i = \sigma_i(\bar{b}_A - \bar{b}_B) + \bar{b}_B = \sigma_i \cdot \Delta\bar{b} + \bar{b}_B$ and the cross section becomes

$$\frac{d\Sigma}{d\Omega}(Q) = \frac{\Delta\bar{b}^2}{V_S}\sum_{ij}^{N}\langle\sigma_i\sigma_j \exp(i\underline{Q}\underline{r}_{ij})\rangle = K \cdot S(Q) \tag{5}$$

With the contrast factor $K = \Delta\rho^2/N_A$ determined from the difference of the coherent scattering length density $\rho_{A,B} = \sum b_j/\Omega$ of both monomers and Avogadro's number N_A. (b_j is the coherent scattering length of atom j and Ω the volume of both a lattice site and a monomer). The definition of Eq. 4 implies an incompressible melt as no free volume has been considered. The structure factor $S(Q)$ represents an average over all chain conformations as indicated by the brackets in Eq. 5; it is defined in units of a molar volume [$\mathrm{cm}^3/\mathrm{mol}$] and is identical to the partial structure factor $S_{AA}(Q)$ determined by intra- and intermolecular correlation between the monomers of polymer A. A separation of $S_{AA}(Q)$ into intra- $\{P(Q)\}$ and intermolecular $\{W(Q)\}$ interference gives

$$S(Q) = S_{AA}(Q) = \Phi V_P \left[P(Q) + \Phi\, n_A W(Q)\right] \quad (6)$$

with the polymer molar volume V_P and the volume fraction Φ of polymer A [6]. The structure factor can be simplified to

$$S(Q) = \Phi(1-\Phi)\; V_P P_{Debye}(Q) \quad (7)$$

which means that $S(Q)$ is determined from the polymer form factor $P(Q)$ alone. This can be understood from the Babinet principle and more easily from the following argument. When $\Phi = 1$, one has a pure melt of A polymers which in the incompressible limit does not scatter, e.g. $S(Q) = 0$ in Eq. 6. Therefore $P(Q) = -n_A W(Q)$ must hold. Moreover, the Debye form factor $P_{Debye}(Q) = 2[x - 1 + \exp(-x)]/x^2$ with $x = R_g^2 Q^2$ (R_g radius of gyration) describes the conformation of polymers in melts to a good approximation as the excluded volume interaction is screened out by other chains and the conformation becomes that of a freely jointed chain [3, 4].

The structure factor in Eq. 7 has a maximum at $Q = 0$ whose magnitude is determined by the polymer molar volume V_P and the composition Φ of the A/B components. The shape of $S(Q)$ is determined by the radius of gyration R_g. In analyzing experimental data one often prefers approximate forms, namely $P_{Debye}^{-1} \cong 1 + (R_g Q)^2/3$ (Zimm approximation) and $P_{Debye} \propto 1/(Q \cdot R_g)^2$ that are valid at $Q < 1/R_g$ and $Q > 1/R_g$, respectively. For Q larger than $1/R_g$ one realizes a scaling law with the so-called fractal dimension $D = 2$. An experimental verification is demonstrated in Fig. 3 showing a 50% isotopic mixture of protonated and deuterated polystyrene [61]. The double logarithmic presentation of the data shows a -2 power law over a large Q interval which corresponds to the expected dimension being equivalent to the exponent $\nu = 1/2$ of the freely jointed chain.

Two more comments have to be given: (1) We neglected the monomer structure ($P_m(Q) = 1$) as its size is too small for the available Q range of SANS. (2) In $d\Sigma/d\Omega$ of Eq. 5 a constant contribution from all monomers, namely $\bar{b}_i = \bar{b}_B$ was neglected as their integration over the sample volume gives a $\delta(Q)$ like contribution to the scattering. This scattering from an order

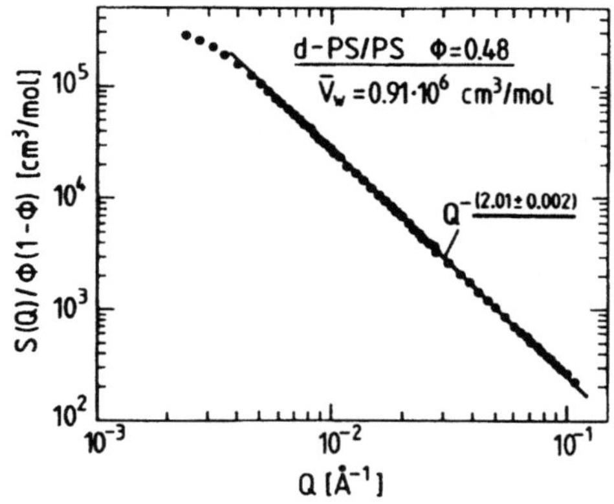

Fig. 3 Structure factor $S(Q)$ versus momentum transfer Q of a 48% isotopic mixture of polystyrene in double logarithmic representation. The Q^{-2} power law at large Q shows that the polymer conformation in melts is that of Gaussian freely jointed chains

of cm large sample would be visible in a Q range of the order of 10^{-8} Å^{-1}. The smallest available Q with neutrons is of the order of 10^{-5} Å^{-1} by the double crystal diffractometer technique [58]. This means that this $\delta(Q)$ contribution can safely be ignored.

2.3 Neutron Scattering Contrast

The contrast factor $K = (\rho_A - \rho_B)^2/N_A = \Delta\rho^2/N_A$ in Eq. 5 describes the interaction between the neutrons and the atoms within the sample which must be sufficiently large for a strong scattering signal. The coherent scattering lengths had to be determined experimentally, and their values can be found in several publications, e.g., in ref. [62]. In polymer research the very different scattering lengths of deuterium and hydrogen, being $b_H = -3.739$ and $b_D = 6.671$ in units of 10^{-13} cm, are of special importance. This leads to strong differences of the scattering length densities of deuterated and protonated polymers as listed in Table 1. Mixing two (A/B) polymers their $\Delta\rho^2$ can be nearly zero when both components are deuterated or protonated as for PB/PVE or d-PB/d-PI, and $\Delta\rho^2$ is usually very large if only one of the components is deuterated. A further advantage of choosing deuterated polymer is the much lower "background" from incoherent scattering in comparison with the protonated ones.

Another possibility is to adjust the coherent scattering lengths of two chemically different polymers in order to make them indistinguishable for SANS studies. In this way individual polymer chains in more complex mixtures can be explored. An example may be polystyrene PS and polybutadiene

Table 1 Coherent scattering length of several polymers and the corresponding contrast factor between them

System	ρ $[10^{10}\ \mathrm{cm}^{-2}]$	$\Delta\rho^2[10^{20}\ \mathrm{cm}^{-4}] \rightarrow$ PB	d-PB	PVE	dPVE	PI	dPI	PS	dPS	PVME
PB	0.416	0	40.3	0	38.7	0.02	43.2	1	36.8	0.004
dPB	6.76		0	40.3	0.014	42.1	0.053	28.5	0.08	41
PVE	0.414		–	0	38.8	0.021	43.2	1	36.8	0.004
dPVE	6.64		–	–	0	40.6	0.122	27.2	0.026	39.6
PI	0.27		–	–	–	0	31	1.32	38.6	0.006
dPI	6.99		–	–	–	–	0	31	0.26	44
PS	1.42		–	–	–	–	–	0	25.6	1.14
dPS	6.48		–	–	–	–	–	–	0	37.6
PVME	0.35		–	–	–	–	–	–	–	0
$C_6H_4Cl_2$	2.36	–	–	–	–	–	–	–	–	–

PB which are indistinguishable for neutrons if a fully deuterated PS is mixed with a partially deuterated PB consisting statistically of 4.4% protonated and 95.6% deuterated monomers.

3
Thermodynamic Model for Polymer Blends

An "ideal" mixture of two chemically distinct polymers represents a model which is not fulfilled in reality. Even mixtures of polymers which are distinguished only by their degree of deuteration or their microstructure such as the vinyl content in polybutadiene (PB) show a small but finite energy of mixing which can easily be measured and also can lead to phase separation. The reason is that the conformational entropy of mixing is inversely proportional to the polymer molar volumes and can therefore be made sufficiently small and properly adjusted to be comparable to the enthalpy of mixing.

3.1
Flory–Huggins Theory

A schematic phase diagram of a symmetrical binary mixture is shown in Fig. 1 in a temperature versus composition representation. A symmetric polymer blend is characterized by two polymer components of the same molar volume and, therefore, with a 50% critical composition. Within mean field approximation the binodal and spinodal phase boundaries of a binary (A/B) incompressible polymer mixture are described by the Gibbs free energy of mixing ΔG according to

$$\Delta G/RT = \frac{\Phi}{V_{\mathrm{A}}} \ln \Phi + \frac{(1-\Phi)}{V_{\mathrm{B}}} \ln(1-\Phi) + \Phi(1-\Phi)\chi \tag{8}$$

as was originally formulated by Flory and Huggins [2]. The first two terms describe the combinatorial entropy of mixing, while the last term represents a free energy of mixing involving the FH parameter χ. Originally, χ was defined being proportional to an enthalpic term χ_{h} according to $\chi = \chi_{\mathrm{h}}/T$; later this parameter had to be extended to a free energy of mixing $\chi = \chi_{\mathrm{h}}/T - \chi_\sigma$ by introducing an empirical non-combinatorial entropic term χ_σ. The necessary extension of the FH parameter to a free energy of mixing was noted by Flory [63] and later mainly forced by SANS experiments. ΔG in Eq. 8 is normalized by the gas constant R and the absolute temperature T. The combinatorial entropic part is inversely proportional to the polymer molar volumes V_{A} and V_{B}. The FH parameter depends on molar volume due to chain-end effects that introduce an additional $1/V$ term and that can be considered as a segmental quantity in the limit of large molar volume [4, 64]. This all means that the phase boundaries of many polymer blends can easily be shifted into an experimentally accessible temperature range by a proper choice of the molar volume.

3.2 Random Phase Approximation

The structure factor $S(Q)$ of an incompressible non-ideal mixture of two polymers with an enthalpy of mixing χ has been derived by deGennes [3] within the random phase approximation (RPA) according to

$$S^{-1}(Q) = \frac{1}{\Phi V_A P^A_{Debye}(Q)} + \frac{1}{(1-\Phi) V_B P^B_{Debye}(Q)} - 2\chi \tag{9}$$

The RPA is a mean field approximation that neglects contributions from thermal composition fluctuations and that assumes the chain conformations to be unperturbed Gaussian chains. The last assumption becomes visible from the Debye form factor in the first two terms, which for $V_A = V_B$ are in accordance with Eq. 7, while the third term involves the FH interaction parameter.

The inverse structure factor at $Q = 0$ is a susceptibility and is related to the Gibbs free energy of mixing ΔG according to

$$S^{-1}(0) = \partial^2_\Phi (\Delta G/RT) \tag{10}$$

Equation 10 is a general result based on the fluctuation-dissipation theorem [3, 4, 9]. So the susceptibility can independently be evaluated from ΔG which in the case of the mean field FH theory leads to a similar result as the RPA,

$$S^{-1}(0) = 2[\chi_S - \Gamma] = (2\Gamma_h/T_S)\frac{(1/T_S - 1/T)}{1/T_S} = C^{-1}_{MF}\tau \tag{11}$$

with the critical amplitude $C_{MF} = T_S/2\Gamma_h$, the reduced temperature $\tau = [1/T_S - 1/T]/1/T_S$, and the FH parameter at the spinodal $2\chi_S = [\Phi V_A]^{-1} + [(1-\Phi)V_B]^{-1}$ which at the critical point becomes $\chi_C = 2/V$ for $V = V_A = V_B$ and $\Phi_C = 0.5$. The susceptibilities $S(0)$ in Eq. 11 and those derived from the RPA (Eq. 8) are identical only if the FH parameter is not dependent on composition. We therefore redefine the FH parameter in the context of SANS experiments according to $2\Gamma = \partial^2_\Phi [\Phi(1-\Phi)\chi]$ as Eq. 10 is the more basic equation. This leads to an effective FH parameter $\Gamma = \Gamma_h/T - \Gamma_\sigma$ with the enthalpic and entropic terms Γ_h and Γ_σ, respectively, while $\Gamma_C \equiv \chi_C = 2/V$ remains the same.

For small Q ($< 1/R_g$) the structure factor in Eq. 9 can be expanded in Zimm approximation according to

$$S^{-1}(Q) = S^{-1}(0) + L_2 Q^2 \tag{12}$$

with $L_2 = [R^2_{g,A}/\Phi V_A + R^2_{g,B}/(1-\Phi)V_B]/3$ determined by the ratios R^2_g/V for the two components and describing the square of the chain statistical segment

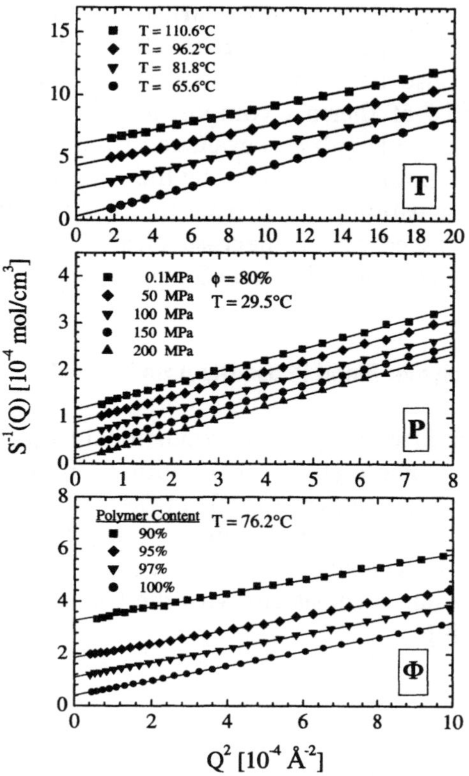

Fig. 4 Structure factor of binary blend dPB/PS in Zimm representation at different temperatures (*top*) and pressures (*middle*), and polymer concentrations (*bottom*), with the other parameters constant. From the fitted *straight lines* the susceptibility and correlation length is evaluated. The increase of scattering is caused by stronger thermal composition fluctuations when approaching the critical point

length. An experimental verification of Eq. 12 is shown in Fig. 4 for a blend of deuterated polybutadiene and protonated polystyrene of about 2000 cm^3/mol molar volume and critical composition described by sample 2 in Table 2. The structure factors are depicted in a Zimm representation of $S^{-1}(Q)$ versus Q^2. All scattering curves follow a straight line in the given Q range in accordance with Eq. 12. The scattering is shown as a function of three separate parameters. The top figures show $S(Q)$ at four temperatures; decreasing the temperature leads to stronger scattering because thermal fluctuations increase when approaching the critical temperature. The middle figures show the sample for different pressures; at higher pressures the scattering becomes stronger because the critical temperature is shifting to higher values and therefore is approaching the sample temperature of 29.5 °C. The lower figures show the blend at constant external conditions, however with different concentrations of a non-selective solvent as specified by the total polymer content; lower scattering is observed for larger solvent content as the solvent leads to a larger miscibility and therefore to a lower critical temperature.

Equation 12 can also be considered as an Ornstein–Zernicke equation describing the degree of thermal composition fluctuations of correlation length ξ. The correlation length ξ is evaluated from $\sqrt{L_2 \cdot S(0)}$ and becomes infinite at the critical point as described by the scaling law $\xi = \xi_0^{MF} \tau^{-0.5}$. The thermodynamic information derived from the susceptibility $S(Q = 0)$ represents a property averaged over a macroscopically large volume due to the inverse relationship between reciprocal (Q) space and real (r) spaces.

3.3
Determination of a Polymer Blend Phase Diagram by SANS

The inverse susceptibility in Eq. 11 is proportional to the reduced temperature τ and is proportional to $(1/T_S - 1/T)$ with T_S the spinodal temperature that becomes the critical temperature T_C in the case of the critical concentration. This means that the inverse susceptibility $S^{-1}(0)$ is positive in the one-phase regime and becomes zero at the spinodal as well as at the critical temperature. This result is consistent with the Gibbs conditions of stability which according to [9]

$$\partial^2_{\Phi}\left(\Delta G/RT\right) \begin{cases} > 0 & \text{Stable} \\ = 0 & \text{Spinodal} \\ < 0 & \text{Unstable} \end{cases} \tag{13}$$

determines the thermodynamic conditions of stability as well as the spinodal border line and the critical point. In this respect the susceptibility in Eq. 10 is a measure of the stability conditions.

An experimental example is given in Fig. 5 where $S^{-1}(0)$ is plotted versus $1/T$, and in Fig. 5b the corresponding phase diagrams of polystyrene (PS) and deuterated polybutadiene (dPB) with different vinyl contents are depicted (Samples 2–4 in Table 2). $S^{-1}(0)$ was measured in a sample with off-critical composition; within the homogeneous regime $S^{-1}(0)$ follows a straight line dependence on τ in agreement with Eq. 11 and extrapolates to $S^{-1}(0) = 0$ at the spinodal temperature T_S. Slightly above T_S the scattering shows strong deviations from the straight line; the neutron intensity abruptly decreases on further lowering the temperature. This behavior is understood from phase decomposition below the binodal line. Below the binodal the precipitated domains grow very fast to sizes of the order of μm length as can easily be visualized from the turbidity of the system. Scattering from the shape of these domains occurs in a Q range that is too small for conventional SANS diffractometers. The observed scattering below T_B therefore measures thermal composition fluctuations within the domains whose strength decreases with diminishing temperature as the distance from the critical temperature

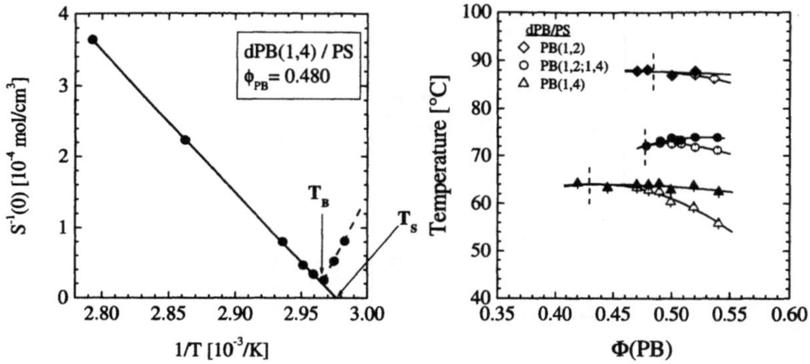

Fig. 5 *Left*: Inverse susceptibility $S^{-1}(0)$ versus inverse temperature for an off-critical blend. The extrapolated *straight line* gives the spinodal temperature, while the bend yields the binodal temperature. Below the binodal fluctuations within the domains are visible which decrease with decreasing temperature. *Right*: Phase diagram of three dPB/PS blends of about 2000 molar volume and with different vinyl content of the dPB component. The different miscibilities and positions of the critical point show that the FH parameter depends on the vinyl content

becomes larger. The critical scattering from fluctuations above and below the phase boundary has been extensively studied in ref. [65]. The so determined binodal and spinodal have been plotted in Fig. 5b. The solid and vertical dashed lines represent a guide for the eye and the critical composition, respectively. This type of experiment becomes necessary in order to determine the critical composition in blends for later analysis of crossover experiments.

Some interesting phenomena can be learned already from the phase boundaries in Fig. 5b. The three mixtures of PS/dPB of about 2000 cm^3/mol were mixed with PB components of different vinyl content (Samples 2–4 in Table 2) [37]. With increasing vinyl content the compatibility is increased by about 25 K, and the critical composition is shifted to about a 5% larger PB volume fraction. For the blend with the statistical dPB(1,2;1,4) copolymer, the critical temperature does not represent the highest temperature for phase decomposition as usually expected (see Fig. 1). This observation can be interpreted as arising from a statistical distribution of vinyl contents for individual, relatively short PB chains. Therefore, here we have the situation of a multi-component blend because the FH parameter changes with the vinyl content as already became clear from the different positions of the phase boundaries. Those blends are so-called "quasi binary" blends when as in this example the PB copolymer chains are considered as a single component. Similar effects are known from polymer polydispersity where the entropy of mixing is proportional to $1/V$ as has been considered by Koningsveld [8] (see also [19]) who in particular shows that the shape of the binodal is sensitively affected by polydispersity.

Table 2 Polymer blends. The vinyl content of PB(1,4), PB(1,2;1,4), and PB(1,2) is 7, 54, and 96%, respectively

No.	System	V_A [cm^3/mol]	N_A	V_B [cm^3/mol]	N_B	Φ_A [cm^3/mol]	$\langle V \rangle$	$\langle N \rangle$
1	dPS/PS	870 000	9060	950 000	9900	0.48	910 000	9100
2	dPB(1,4)/PS	2200	36.5	1800	18.8	0.42	2215	28.2
3	dPB(1,2;1,4)/PS	2000	33.2			0.505	1900	23.9
4	dPB(1,2)/ PS	2100	34.9			0.5	1940	24.4
5	dPS/PVME	218 000	2270	192 000	3390	0.19	212 500	2422
6	dPB(1,4)/PB(1,2;1,4)	28 000	463.6	20 600	341	0.5	23 700	392
7	dPB(1,4)/PB(1,4)			27 500	455.3	0.5	27 700	458.6
8	dPB(1,2)/PB(1,2;1,4)	20 500	339.4	20 600	341	0.5	20 600	341
9	PPMS/dPS	3960	33.3	4320	43.6	0.5	4130	37.8
10	PEE/PDMS/PEE-PDMS	2010	30.5	2270	29.2	0.516	12 000	168
11	dPB(1,4)/PS/dPB(1,4)-PS	2720	45	2180	22	0.42	15 400	209.2

4
Mean Field to 3D-Ising Crossover in Polymer Blends

Thermal fluctuations can be described within the Gaussian approximation at sufficiently high temperatures above the critical temperature. For these situations, the system fulfills the conditions of mean field approximation [9]. On the other hand, thermal composition fluctuations become strong near the critical temperature, leading to non-linear effects which asymptotically close to the critical temperature imply that the system obeys the universality class of 3D-Ising critical behavior. Thermal fluctuations are described by the Ginzburg–Landau Hamiltonian which is written as a functional of the spatially varying order parameter $\Phi(x)$,

$$H = \int \mathrm{d}^d x \left\{ c_2 \left[\nabla \Phi(x)\right]^2 + r\Phi^2(x) + u\Phi^4(x) \right\} \tag{14}$$

The order parameter $\Phi(x)$ is represented by the volume fraction Φ of one of the two polymer components (see Fig. 1). The first term represents the surface energy and leads to the correlation length of thermal fluctuations; the second term corresponds to the susceptibility according to $r^{-1} \equiv 2S(0)$, and the third term describes non-linear effects from the fluctuation fields $\Phi(x)$ [9]. When the Gaussian approximation is valid, the last term is negligible.

Linear and non-linear effects of thermal composition fluctuations become visible in a scattering experiment. Within the mean field and Ising regimes the susceptibility $S(0)$ and the correlation length ξ are described by simple scaling laws as functions of the reduced temperature τ according to $C\tau^{-\gamma}$ ($\xi_0\tau^{-\nu}$) with the critical amplitudes $C(\xi_0)$ and the critical exponents $\gamma(\nu)$. The critical exponents $\gamma(\nu)$ are known to be equal to $\gamma = 1$ and 1.239 ± 0.003 and $\nu = 0.5$ and 0.634 ± 0.001 in the mean field and Ising cases, respectively [66]. The mean field case has already been discussed in the context with Eq. 11.

Mean field and Ising behavior are observed in the limits where the system is asymptotically far and close to the critical temperature, respectively. A further theoretical approach by so-called crossover functions is needed in order to describe the susceptibility and correlation length over the full miscible range of the blend and thereby to bridge both asymptotic limits. Such crossover functions and the corresponding SANS results from polymer blends under various conditions will be discussed in the following sections.

4.1
Theoretical Background

To a large extent, the formalism of crossover functions has been developed by Sengers and coworkers [10] in recent years. It has been shown that the crossover from mean field to Ising behavior could be non-universal and rather complex. In the asymptotic critical Ising type regime the correlation length ξ is the only relevant length, while in the crossover regime an additional microscopic length could become relevant and enforce a non-universal and non-homogeneous crossover behavior [29]. Such a non-universal crossover behavior has been observed in dilute polymer solutions in which the size of the polymer was quite naturally interpreted as the microscopic length scale [67]. In most polymer blends so far, the variation of $S(0)$ can be described by a universal and continuous crossover function that is solely determined by a crossover temperature. Only in a few cases does the radius of gyrations of the chains seem to appear as a microscopic length that is visible in the crossover behavior [68]. This means that a relevant microscopic length scale usually does not appear in polymer blends and can therefore be neglected. A corresponding crossover formalism for the susceptibility and correlation length has been derived from a more general formalism as

$$S^{-1}(0) = (1/C_{MF})\,\tau\, Y^{(\gamma-1)/\Delta}\left[1+\frac{u^*\nu}{\Delta}\left(\frac{1}{2\Delta}+\frac{Y}{1-Y}\right)^{-1}\right] \tag{15}$$

$$\xi = \xi_0^{MF}\,\tau^{-1/2}Y^{-(2\nu-1)/2\Delta}\,. \tag{16}$$

Upon taking the cut off wave number $\Lambda \to \infty$ which corresponds to a vanishing microscopic cut-off length [10, 30]. The susceptibility and correlation length are functions of the reduced temperature τ and the crossover function Y. The critical exponents γ and ν are those of the 3D-Ising model and $\Delta = 0.54 \pm 0.03$ is a universal correction-to-scaling exponent [66]. The crossover function Y is either given as an implicit function of the inverse correlation length $\kappa = \xi_0^{MF}/\xi$ or of the reduced temperature τ according to $1 - Y = (\bar{u}\Lambda/\kappa)\,Y^{\nu/\Delta}$ or $(1 - Y)^2 Y^{-1/\Delta} = \bar{u}^2\Lambda^2/\tau$, respectively. A vanishing microscopic length ($\Lambda \to \infty$) also means that the reduced coupling constant $\bar{u} \to 0$ because of its definition $\bar{u} = u/u^*\Lambda$. The parameters u and $u^* \cong 0.472$ represent the coupling constant in the Ginzburg–Landau Hamiltonian in Eq. 14 and the RG fixed-point coupling constant, respectively. For the mean field approach, using $\bar{u}^2\Lambda^2/\tau \ll 1$ and $Y = 1$ gives the scaling laws $S(0) = C_{MF}\tau^{-1}$ and $\xi = \xi_0^{MF}\tau^{-1/2}$ with the corresponding mean field critical amplitudes C_{MF} and ξ_0^{MF} in conformity with the expression in Eq. 11. Within the asymptotic regime of 3D-Ising behavior one finds $\tau \to 0$ and $Y \to 0$ and therefore $Y = (\bar{u}^2\Lambda^2/\tau)^{-\Delta}$. From Eqs. 15 and 16 and using the proper Y one gets $S(0) = C_+\tau^{-\gamma}$ and $\xi = \xi_0\tau^{-\nu}$ gives the asymptotic scaling laws of the 3D-Ising model. The ratios of the Ising and mean field critical amplitudes are given as

$$(C_+/C_{MF})^{1/(\gamma-1)} = \bar{u}^2\Lambda^2 \qquad \left(\xi_0/\xi_0^{MF}\right)^{2/(2\nu-1)} = \bar{u}^2\Lambda^2 \tag{17}$$

In the present case of a zero microscopic length ($\Lambda \to \infty$) the crossover between the two universality classes is determined by a single parameter, namely $(\bar{u}\Lambda)^2$ which is proportional to the Ginzburg number Gi and thereby to the crossover temperature T_X (see Fig. 1). In conformity with our earlier work [31] we define the Ginzburg number as $Gi = 0.069\,\bar{u}^2 \Lambda^2$ and finally obtain in Eq. 18

$$Gi = 0.069\,(C_+/C_{\mathrm{MF}})^{1/(\gamma-1)} = 0.069\,\left(\xi_0/\xi_0^{\mathrm{MF}}\right)^{2/(2\nu-1)} \tag{18}$$

the Ginzburg number represented in terms of the ratio of the Ising and mean field critical amplitudes and a large exponent $1/(\gamma-1)$ $[2/(2\nu-1)]$ of about 4 as determined from the Ising critical exponents. Quite generally, Gi can be identified with the reduced temperature at which the difference between the Ising approximation and the crossover function reaches 10%.

Another form of crossover function for the susceptibility

$$\hat{t} = \left(1 + 2.333\widehat{S}(0)^{\Delta/\gamma}\right)^{(\gamma-1)/\Delta} \times \left[\widehat{S}^{-1}(0) + \left(1 + 2.333\widehat{S}(0)^{\Delta/\gamma}\right)^{-\gamma/\Delta}\right] \tag{19}$$

has been derived by Belyakov and Kiselev from a renormalization group approach and an ε-expansion [27]. In this expression the rescaled reduced temperature $\hat{t} = \tau/Gi$ is an explicit function of the rescaled susceptibility $\widehat{S}(0) = S(0)\,Gi/C_{\mathrm{MF}}$, the critical exponents γ and Δ are from the 3D-Ising model. So, the parameters Gi, C_{MF}, and T_{C} are the parameters of the susceptibility that are fit to the experimental SANS data. Most of the SANS susceptibility data so far were analyzed with Eq. 19 as it was the only available and proper crossover function at the time when the experiments were performed. But in any case both crossover functions describe a similar situation as in both cases a microscopic length scale was neglected.

The expressions for the critical amplitudes are given for the mean field case in Eq. 11. For the Ising case they are expressed in terms of the molecular volumes for symmetric blends according to $C_+ \propto V^{(2-\gamma)}$ and $\xi_0 \propto V^{(1-\nu)}$ as derived by Binder [4, 16]. These expressions lead to the representations

$$Gi \propto \left[V^{(2-\gamma)}(2/V + \Gamma_\sigma)\right]^{1/(\gamma-1)} \qquad Gi \propto \left[V^{(2-2\nu)}(2/V + \Gamma_\sigma)\right]^{1/(2\nu-1)} \tag{20}$$

and in case of a zero entropic term Γ_σ to the universal scaling law $Gi \propto 1/V$. The last expression was originally derived by deGennes assuming an incompressible polymer melt [3]. It becomes apparent from the large exponents $1/(\gamma-1) \cong 4.23$ and $1/(2\nu-1) \cong 3.73$ that a positive finite entropic term Γ_σ strongly enhances Gi and thereby the effect of thermal composition fluctuations near the critical temperature.

The above expressions for Gi are valid for UCST systems showing phase decomposition at low temperatures. These systems are characterized by (Γ_{h} and Γ_σ) > 0 and $|\Gamma_\sigma| < \Gamma_{\mathrm{C}}$ if $\Gamma_\sigma < 0$ as otherwise no miscibility is achieved. There are several blends, showing phase decomposition at high temperatures, and which are so-called LCST systems. The FH parameter of such blends have to fulfill the following conditions, namely (Γ_σ and Γ_σ) < 0 and $|\Gamma_\sigma| > \Gamma_{\mathrm{C}} = 2/V$, in order to show phase transitions at a finite critical temperature. The corresponding critical amplitude for the mean field susceptibility becomes $C_{\mathrm{MF}}^{-1} = 2\left[-\left(2/V + \Gamma_\sigma\right)\right]$ because of the necessary redefinition of the reduced temperature according to $\tau \propto (1/T - 1/T_C)$. For a positive critical amplitude and thereby positive susceptibility the above conditions for a finite T_C have to be fulfilled, and the corresponding Ginzburg criterion for $S(0)$ is expressed as

$$Gi \propto \left[V^{(2-\gamma)}(-2/V + |\Gamma_\sigma|)\right]^{1/(\gamma-1)} \tag{20a}$$

This means that in polymer blends quite large values of Gi are possible and also implies a general need for a more sophisticated analysis of the scattering data using Eqs. 15,16 or 19. This also means that the FH parameter has to be evaluated from the mean field critical amplitudes C_{MF} in Eq. 11 and from the "mean field" critical temperature T_C^{MF} which is approximately related to the "real" critical temperature T_C according to $T_C^{MF} = T_C(1 + Gi)$ [10, 28]. T_C is lower than T_C^{MF} by several degrees Kelvin because of the stabilization effect of thermal fluctuations.

The dependence of Gi in Eq. 20 on the entropic term Γ_σ leads to a connection with other theoretical approaches to Gi for polymer blends. According to equation of states theories a finite entropic term Γ_σ arises from the sample compressibility [33] which also becomes apparent from the Lattice Cluster Theory (LCT) [12]. It has to be mentioned here that the Ginzburg criterion derived on the basis of mean field parameters by Bates et al. [20] and Hair et al. [21] behave inversely proportional to the non-combinatorial entropy term Γ_σ [see also Eq. (4.2) in Dudowicz et al. [12]]. So there still seems to exist some confusion with respect to the effect of Γ_σ on Gi.

4.2 Binary Homopolymer Blends

A first experimental example exhibiting a crossover from mean field to Ising critical behavior is depicted in Fig. 6 showing the inverse susceptibility and the square of the inverse correlation length versus inverse temperature for a blend of high molar volume deuterated polystyrene (dPS) and poly(vinyl methylether) (PVME) with the critical composition (Sample 5 in Table 2). The susceptibility and correlation length presented were analyzed with the asymptotic crossover function of Eqs. 19 and 16, respectively, and the best fits are shown by the solid lines; the agreement between experiments and theory is excellent. The experimental conditions for determining the correlation length corresponds to those of the susceptibility giving the same Ginzburg number $Gi = (4.5 \pm 0.5)10^{-5}$ and roughly a 0.5 K shift of the critical temperature from the mean field estimate. The correlation length shows qualitatively the same behavior as $S(0)$ and leads to an additional critical amplitude parameter. From the fit of parameters, the critical temperature, the Ginzburg number and the mean field and 3D-Ising critical amplitudes were derived from which the FH parameter were evaluated. The positive slope of $S^{-1}(0)$ in

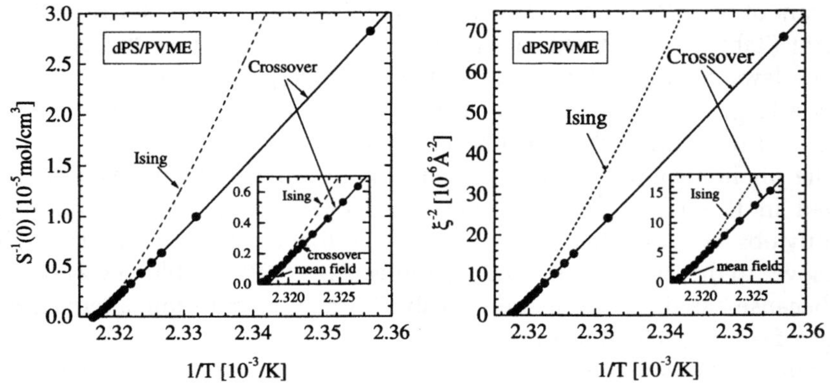

Fig. 6 Inverse susceptibility $S^{-1}(0)$ and inverse square of the correlation length versus inverse temperature for the critical LCST high molar volume blend dPS/PVME. The *solid lines* represent a fit of the corresponding crossover functions. The mean field approximation is visible in the *insets*, the Ising behavior by the *dashed lines*

Fig. 6 is related to a negative enthalpic Γ_h (Eq. 11). In the case of a negative Γ_h, the PS and PVME heterocontacts are preferred. Only in the case of a negative entropy of mixing ($\Gamma_\sigma < 0$) that is larger than the always positive combinatorial entropy of mixing does a positive driving force for phase separation exist. The critical temperature is defined to occur at $S^{-1}(0) = 0$ which corresponds to an infinite correlation length. The critical condition of $S^{-1}(0) = 0$ is very well approached, indicating that the experiments were performed at the critical composition.

Figure 7 depicts the susceptibility of two more conventional UCST blends of critical concentration and about 2000 cm^3/mol molar volume. The phase

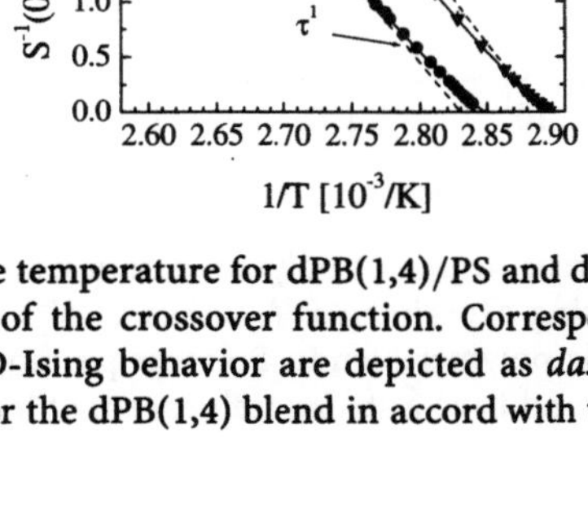

Fig. 7 $S^{-1}(0)$ versus inverse temperature for dPB(1,4)/PS and dPB(1,4;1,2)/PS blends. The *solid lines* represent a fit of the crossover function. Corresponding asymptotic scaling laws of mean field and 3D-Ising behavior are depicted as *dashed lines*. A lower critical temperature is observed for the dPB(1,4) blend in accord with the phase diagram in Fig. 5

diagram of two dPB/PS blends with 7 and 54% vinyl contents for the PB component (Table 2) are shown in Fig. 5b. The dashed lines give the asymptotic scaling laws corresponding to mean field and Ising behaviors. These blends show a larger crossover range of the order of 13 K in comparison with the range of 1 K for the high molecular dPS/PVME blend.

The phenomenon of crossover behavior in polymer blends was first observed in this large molar volume dPS/PVME blend [18, 19] and subsequently observed by other groups [20–26]. At that time the observation of the crossover to the Ising case in the high molecular PS/PVME blend was surprising because based on the "incompressible" Ginzburg criterion, a temperature range of roughly 0.05 K was expected.

Figure 8 shows a collection of Ginzburg numbers from various polymer blends versus the degree of polymerization [31]. A strong variation of *Gi* over four orders of magnitude is obtained for samples ranging over two orders of magnitude in *N*. The points lie within an area described by N^{-1} and N^{-2} scaling laws, the first of which was proposed by deGennes [3] and the second by SANS experiments. A remarkable observation is related to the order of magnitude larger *Gi* that is found for blends composed of shorter chains between $N = 10$ and 100 as compared with the $Gi \cong 10^{-2}$ of low molecular liquids. These observations in the context of Eq. 20 clearly demonstrate that the entropic term Γ_σ and thereby the compressibility plays an important role in polymer mixtures and is responsible for the non-universal behavior of the variation of *Gi* with *N*. These observations also must be considered with the understanding that the degree of polymerization of ordinary UCST polymer

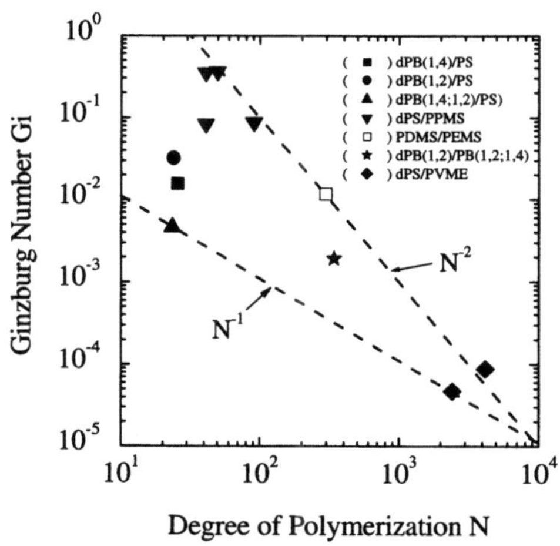

Fig. 8 Ginzburg number versus degree of polymerization from different polymer blends as determined from scattering experiments. A universal N^{-1} scaling behavior has been predicted for incompressible polymer blends. Polymer packing effects enter through the non-combinatorial entropy and have a strong influence that leads to an increase of *Gi*

blends with at least partial miscibility are usually not larger than of the order of $N = 100$.

4.3 Binary Blends of Statistical Copolymers

Statistical copolymers often show the interesting effect of better compatibility; comparing polymers of similar length, one may find a "C" homopolymer miscible with an "A-B" statistical copolymer even though the "A, B, and C" homopolymers are not miscible with each other. This phenomenon has been interpreted by the interplay of the intermolecular interactions between the A-C and B-C monomers and the intramolecular interaction between the A-B monomers reducing the total enthalpy of mixing [69, 70]. While this explanation seems to be sufficient to interpret an enhanced compatibility determined from cloud point experiments, it completely failed to correctly describe recent SANS experiments on binary blends of PS and PB with different vinyl contents [37]. For a better understanding one needs a full analysis of the FH parameter including the enthalpic and non-combinatorial part of a free energy of mixing which cannot separately be determined from cloud point measurements alone.

In this paragraph we will discuss three blends of dPB(1,2)$_x$-dPB(1,4)$_{(1-x)}$/PB(1,2)$_y$-PB(1,4)$_{(1-y)}$ which consist of the same PB statistical copolymer with different vinyl content x and y of the deuterated and protonated component (Sample 6–8 in Table 2). These systems represent the simplest realization of copolymer blends of the general type A_x-$B_{(1-x)}$/C_y-$D_{(1-y)}$ with x and y the volume fraction of the A and C monomers, respectively. The A and B monomers were deuterated in order to have good neutron scattering contrast. Symmetric mixtures with chains of similar molar volume between 2 to 28k ($k \equiv 10^3$ cm^3/mol) and with x and y of 0.07, 0.54, 0.91 were recently published in ref. [64]. Figure 9 depicts the susceptibility versus inverse temperature $1/T$ of the larger molar volume blends, namely dPB(1,2)/PB(1,2;1,4) and dPB(1,4)/PB(1,2;1,4) together with that of the isotopic mixture of PB(1,4). The scattering from the PB(1,4) isotopic mixture is rather weak as expected, while the other two samples give very different results in spite of their similar molar volumes and symmetries with respect to their microstructure. At first glance, this strong difference is very surprising as the classical theories of copolymers predict for $|x - y| \cong 0.5$ the same FH parameter and therefore the same degree of scattering. The blend containing dPB(1,2) and PB(1,2;1,4) shows large scattering with $S^{-1}(0) = 0$ at the critical temperature of $T_C = 69$ °C and, furthermore, a crossover behavior according to Eq. 15 and Eq. 19 as depicted by the solid line in Fig. 9. A Ginzburg number of $Gi = (1.9 \pm 0.8)10^{-3}$ was evaluated and is depicted by the symbol ⋆ in Fig. 8. The susceptibility of the other two blends in Fig. 9 were fitted by straight lines in accordance with the mean field approximation of Eq. 11.

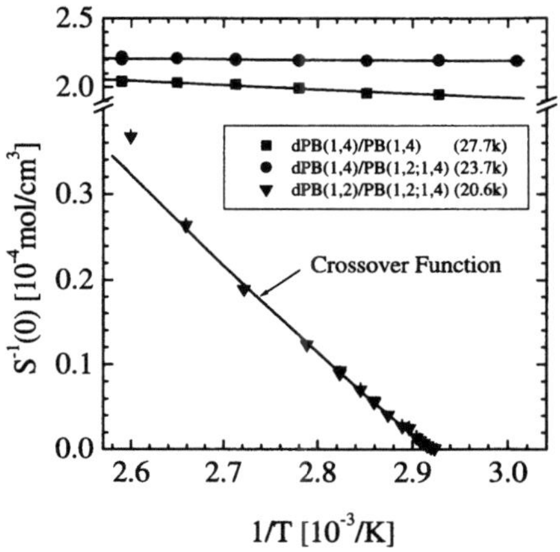

Fig. 9 Inverse susceptibility $S^{-1}(0)$ versus inverse temperature for three dPB/PB statistical copolymer blends with different vinyl contents and molar volumes as indicated. All blends are of 50% composition. A remarkably strong scattering is found for the dPB(1,2)/PB(1,2;1,4) sample that is described by a crossover function in contrast to the weak scattering from dPB(1,4)/PB(1,2;1,4). The overall decrease of the interaction energy on deuteration explains this observation

A first observation can be established, namely, that blends with deuterated PB(1,2) show a much stronger tendency towards incompatibility than blends with deuterated PB(1,4). This observation is supported by the finding that a corresponding mixture of dPB(1,2) and PB(1,4) was not miscible below 200 °C.

4.3.1
Theory

Quite recently, Dudowicz and Freed formulated an expression for the FH parameter for statistical copolymers within their "Lattice Cluster Theory for Pedestrians" (see Eq. 13 in [71] or Eq. 1 in [72]). This expression leads to the classical theory of

$$\begin{aligned} \Gamma \equiv & \Gamma_{\mathrm{h}}(\mathrm{AC})xy + \Gamma_{\mathrm{h}}(\mathrm{BC})(1-x)y + \Gamma_{\mathrm{h}}(\mathrm{AD})x(1-y) \\ & + \Gamma_{\mathrm{h}}(\mathrm{BD})(1-x)(1-y) - \Gamma_{\mathrm{h}}(\mathrm{AB})x(1-x) - \Gamma_{\mathrm{h}}(\mathrm{CD})y(1-y) \end{aligned} \quad (21)$$

only if Γ_σ is neglected and the monomers are taken to occupy single lattice sites [69, 70]. In the case of a positive intramolecular $\Gamma_{\mathrm{h}}(\mathrm{AB})$ and $\Gamma_{\mathrm{h}}(\mathrm{CD})$, the overall enthalpic term in Eq. 21 is reduced with a resultant better compatibility between the polymers in the blend.

The enthalpy of the deuterated $\mathrm{dPB}(1{,}2)_x\mathrm{dPB}(1{,}4)_{(1-x)}$ and protonated $\mathrm{PB}(1{,}2)_y\mathrm{PB}(1{,}4)_{(1-y)}$ are determined by the nearest neighbor attractive van der Waals energies $\varepsilon_{\mathrm{CC}}$, $\varepsilon_{\mathrm{DD}}$, and $\varepsilon_{\mathrm{CD}}$, respectively, describing the interaction between the protonated 1,2–1,2, 1,4–1,4, and 1,2–1,4 pairs of monomers. Hence, the enthalpic term, representing the intramolecular interactions such as between the C and D monomers, is determined by the exchange energy $\varepsilon = \varepsilon_{CC} + \varepsilon_{DD} - 2\varepsilon_{CD}$ in the form with $\Gamma_h(CD) = (2z/\Omega)\,\varepsilon$, with z and Ω representing the number of nearest neighbors (theoretical number is $z = 6$) and the monomer volume ($\Omega = 60.4\,\mathrm{cm}^3/\mathrm{mol}$ for PB), respectively. A simplification is achieved if the interaction energy of deuterated monomers is described by the polarizability model of Bates et al. [73], which leads to $\varepsilon_{AA} = \gamma^2\varepsilon_{CC}$, $\varepsilon_{BB} = \gamma^2\varepsilon_{DD}$, $\varepsilon_{AB} = \gamma^2\varepsilon_{CD}$, $\varepsilon_{AC} = \gamma\varepsilon_{CC}$, $\varepsilon_{BD} = \gamma\varepsilon_{DD}$, $\varepsilon_{AD} = \varepsilon_{BC} = \gamma\varepsilon_{CD}$. The value of γ is slightly less than one because of the smaller interaction between deuterated monomers in comparison with the corresponding protonated ones. In the case of symmetric blends with $\Phi_C = 0.5$, one gets

$$\Gamma_h = (2z/\Omega)\left[\alpha^2\,\varepsilon_{CC} + \beta^2\,\varepsilon_{DD} - 2\alpha\beta\,\varepsilon_{CD}\right] \tag{22}$$

With the coefficients $\alpha = (y - \gamma x)$ and $\beta = [\gamma(1-x) - (1-y)]$ (see Eqs. 1 and 11 in [71, 72]). In the case of a fully protonated blend with $\gamma = 1$, one gets $\Gamma_h = (2z/\Omega)\varepsilon\,[x - y]^2$, and for the PB(1,4)/PB(1,2) homopolymer blend this becomes $\Gamma_h = (2z/\Omega)\varepsilon$. For blends with one component deuterated, one gets for ($x = 0$; $y = 1$) and ($x = 1$; $y = 0$), respectively, $\Gamma_h = (2z/\Omega)\,(\varepsilon_{CC} + \gamma^2\varepsilon_{DD} - 2\gamma\varepsilon_{CD})$ and $\Gamma_h = (2z/\Omega)\,(\gamma^2\varepsilon_{CC} + \varepsilon_{DD} - 2\gamma\varepsilon_{CD})$, e.g., in the case of different van der Waals energies ε_{DD} and ε_{CC}, deuteration leads to a different Γ_h.

Lattice Cluster Theory gives in addition an entropic term which for the PB copolymer blends reads

$$\Gamma_\sigma = -(x - y)^2/(4z^2\Omega) \tag{23}$$

This negative non-combinatorial entropy of mixing Γ_σ is related to the structure of the monomers, e.g., depending on the number of lattice sides per monomer that are occupied by tri- and tetrafunctional united atomic units of the (1,2) and (1,4) monomers; compressibility and chain-end effects were not considered [71, 72]. A more detailed description of this theory is found in the review article by Freed and Dudowicz of the same volume.

4.3.2 Experimental Results

The FH parameters Γ_h and Γ_σ for the PB copolymer blends are depicted in Fig. 10 versus the vinyl content of the protonated PB component. The solid and open triangles and spheres represent blends with fixed vinyl content of $x = 0.07$, 0.16, and 0.91 of the deuterated PB component, respectively. The FH parameters were corrected for chain-end effects which are rather strong and are not shown here (see Fig. 6 in [64]).

The blend with $x = 0.91$ (dPB(1,2)) and $y = 0.54$ vinyl contents yields a relatively large enthalpic Γ_h as already evident from the data in Fig. 9; on the

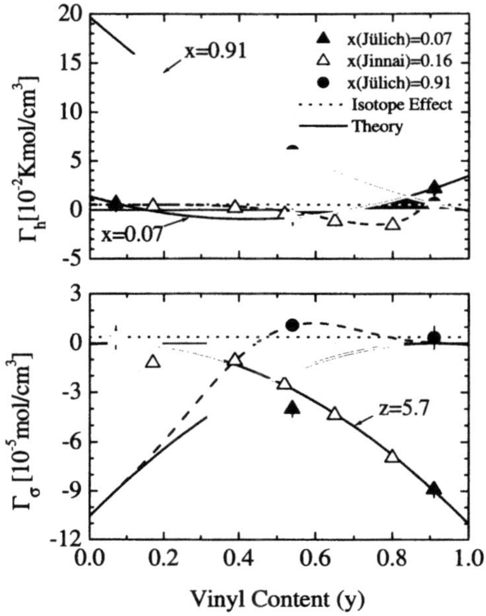

Fig. 10 Enthalpic and entropic FH parameters for dPB/PB statistical copolymer blends in the limit of infinite molar volume as a function of the vinyl content of the deuterated and protonated chains. The enthalpic terms represented in terms of the van der Waals interaction energy and a polarization parameter for deuteration as depicted by the theoretical *solid lines*. The entropic terms are represented in terms of an expression for the non-combinatorial entropy from LCT calculations with a slightly smaller coordination number

other hand, a relatively small Γ_h is observed for the blends with dPB(1,4), for a limited interval of the vinyl content of the protonated component Γ_h even becomes negative. The entropic Γ_σ is negative for the dPB(1,4) blend, while a slightly positive value has been found in the case of the dPB(1,2) blend.

The solid lines in Fig. 10 represent the LCT calculations using Eqs. 21 and 22, while the dashed and dotted lines are a guide for the eye, the latter line showing the isotope effect. The corresponding fit parameters are given in Table 3. Excellent agreement between experiment and theory is achieved, thus explaining the physical reason for the observed quite strong differences due to microstructure and deuteration. The enthalpic Γ_h represented by the solid symbols is fully described by a single set of van der Waals energy parameters ε_{ij} and a single deuteration polarizability factor γ. To make these fits unique, a fixed value of ε_{CC} was taken from PVT measurements of a dPB(1,2) melt [39].

The enthalpic part of the FH parameters appears very much influenced by the isotope effect, e.g., whether the PB(1,2) or PB(1,4) is deuterated or not. If one considers a PB copolymer mixture of only protonated components, using the van der Waals interaction parameters in Table 3 and setting

Table 3 Parameters from fit of copolymer theory

$z = 5.7$
$\varepsilon_{CC} = 211.6\,K^{*)}$
$\varepsilon_{DD} = 250.18\,K$
$\varepsilon_{CD} = 230.56\,K$
$\gamma = 0.988$

*) This number was taken from Frielinghaus et al. as a fixed parameter [39]. ε_{CC}, ε_{DD}, and ε_{CD}, respectively, describe the interaction between the protonated 1,2–1,2,1,4–1,4, and 1,2–1,4 pairs of monomers

$\gamma = 1$, one gets $\Gamma_h = 0.131(x-y)^2$ [Γ_h is always given in units of K mol/cm^3], which is symmetric in x and y, always positive, and zero only for $x = y$, i.e., when the microstructure in both components is the same. A maximum value for $\Gamma_h = 0.13$ is evaluated in the case of $|x-y| = 1$, which corresponds to the PB(1,4)/PB(1,2) blend. If one component is fully deuterated, as described in Eq. 22, one gets very different values for Γ_h, namely, 0.044 and 0.228 for dPB(1,4)/PB(1,2) and dPB(1,2)/PB(1,4), respectively (see also Fig. 10). So, depending on whether PB(1,4) or PB(1,2) is the deuterated component a respective 70% decrease or increase of Γ_h is observed. The smaller Γ_h of the dPB(1,4)/PB(1,2) blend is caused by the 18% larger interaction energy between the (1,4) monomer units (ε_{DD}) in comparison with those between the (1,2) units (ε_{CC}). Hence, a 1.2% change in ε upon deuteration leads to a relatively large change of the enthalpic FH parameter, and, furthermore, in some cases to a transition from an UCST to LCST phase behavior.

The physical basis of the behavior of Γ_h in Fig. 10 becomes more transparent if a simplified version of Eq. 22, namely Eq. 24, is used, which is

$$\Gamma_h = (2z/\Omega)\left[\beta\varepsilon_{DD}^{0.5} - \alpha\varepsilon_{CC}^{0.5}\right]^2 \qquad (24)$$

based on the assumption of dispersive (van der Waals) interactions fulfilling the relationship $\varepsilon_{CD} = \sqrt{\varepsilon_{CC}\varepsilon_{DD}}$. In our case this relationship is fulfilled within 0.2%. The form of Eq. 24 is similar to the relation for FH parameter $\Gamma_h \propto [\beta\Delta_{DD} - \alpha\Delta_{CC}]^2$ in terms of the solubility parameters for the (1,4) (Δ_{DD}) and (1,2) (Δ_{CC}) monomers. The asymmetry effect of deuteration can be more easily understood on the basis of Eq. 24, and the corresponding presentation in Fig. 11. The (1,4) monomers are more stronger bound than the (1,2) monomers as they show a stronger attractive energy and thereby a larger solubility parameter. This difference of solubility parameters is of microstructural origin. A general trend is reported in the literature, namely, the solubility parameter decreases with increasing content of the branched vinyl content [74]. As demonstrated by the dashed and dotted lines in Fig. 11, deuteration of the (1,4) and (1,2) polymers leads to a decrease and increase

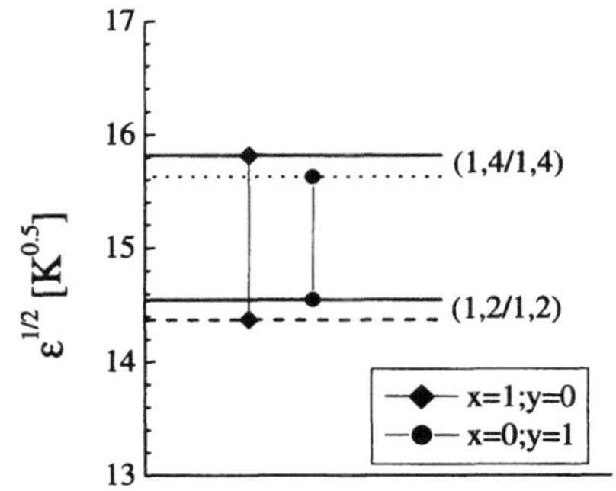

Fig. 11 The square root of the van der Waals interaction energies are proportional to the solubility parameters and are depicted for the (1,4) and (1,2) PB monomers as well as their decrease caused by deuteration. In the case of dispersion forces, the enthalpic term is determined by their difference (*vertical lines*); so deuteration of PB(1,2) and PB(1,4) leads to a larger and a smaller Γ_h, respectively. The different compatibility of PB/PS blends in Fig. 5 can also be concluded from the data in this figure and the larger solubility parameter of PS

of Γ_h, respectively. Similar effects were found in blends of saturated polybutadiene polymers [74].

The expression for entropic Γ_σ in Eq. 23 describes the data with the $x = 0.07$ samples and the results by Jinnai et al. [75] very well assuming $z = 5.7$ nearest neighbors. Because of obvious limitations inherent in any lattice model, the parameter z is allowed to vary freely to compensate for deficiencies of the model. The closeness of $z = 5.7$ to the theoretical $z = 6$ is, however, quite convincing. According to the theory, there seems to be no isotope effect on Γ_σ. Deviations from theory were found for the dPB(1,2)/ PB(1,2;1,4) blend, which shows a positive Γ_σ that cannot be described without possibly including chain semi flexibility effects.

4.4 Polymer Blends in an External Pressure Field

Besides temperature and polymer composition, pressure is a third parameter needed to completely determine the thermodynamic equilibrium state of a binary mixture. So far, only a few systematic SANS studies exist for polymer blends in external pressure fields [34–41]. Those experiments were also performed in our laboratory for which a temperature-pressure cell was developed for in-situ investigations allowing pressure and temperature fields between $0.1 \leq P(\mathrm{MPa}) \leq 200$ and $-20 \leq T(°\mathrm{C}) \leq 200$, respectively. A temperature control better than 0.01 K allowed also precise exploration of thermal composition fluctuations near the critical point [34].

The SANS data near the critical point were analyzed with the corresponding crossover functions (Eqs. 15–19) and the effect of pressure was mainly interpreted in terms of the FH parameter. A theory of compressible blends including the effects of pressure has been incorporated in equation of states theories by Patterson and Robard [33] and included in the LCT by Tang and Freed [76]. A comparison between LCT theory and SANS data for the pressure dependence has been presented in a recent publication by Frielinghaus et al. [39].

4.4.1
SANS Results

As an example Fig. 12 presents the inverse susceptibility of a dPB(1,4)/PS blend (Sample 2 in Table 2) versus the inverse temperature measured in four pressure fields between 0.1 and 150 MPa. The solid lines represent a fit with the crossover function in Eq. 19 and show good agreement. The parameters derived from $S(0)$, namely, the pressure dependent phase boundaries, the Ginzburg number, and the FH parameter will be discussed in the following figures.

The spinodal (critical) and binodal temperatures of the three samples are depicted in Fig. 13 versus pressure. In all samples an increase of the phase boundaries with pressure is observed. Such a behavior is expected because of the reduced free volume. The shape of the phase boundary is linear for the blends with the dPB(1,4) and dPB(1,4;1,2) copolymers and is more parabolic for the blend with dPB(1,2). The dPB(1,2) sample was at the critical composition while the other two samples are slightly off critical composition as seen from the deviation between the spinodals and binodals. The worst compatibility is observed for the dPB(1,2)/PS sample, the best one for the dPB(1,4)/PS, and for the copolymer sample it lies in between.

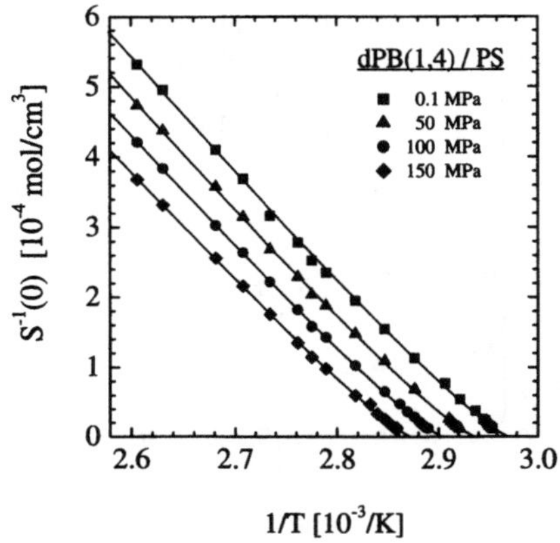

Fig. 12 $S^{-1}(0)$ versus inverse temperature of a dPB(1,4)/PS blend in different pressure fields. Pressure leads to reduced polymer compatibility, and the *solid line* represents a fit of the crossover function

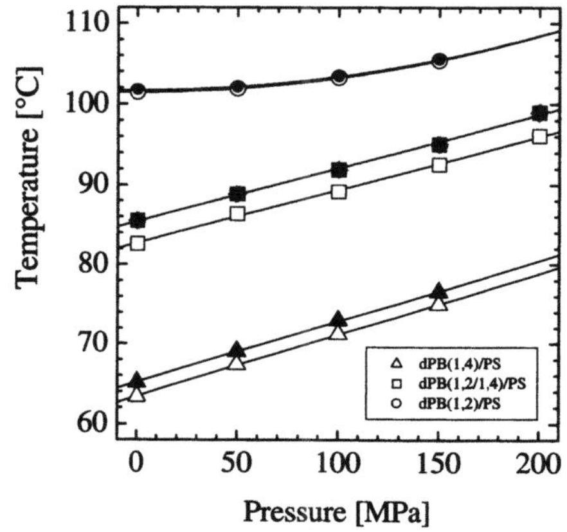

Fig. 13 Binodal and spinodal temperature versus pressure for the three dPB/PS blends as indicated in the *inset*. The phase boundaries of the PB(1,2) blend increase with a parabolic shape, while the variation is linear for the other two blends

The Ginzburg number of the PPMS/PS and PB/PS blends in Fig. 14 always decreases with pressure but shows characteristic differences [36, 37]. The PPMS/PS blend is characterized by a relatively large *Gi* at ambient pressures and a rather fast decrease at larger pressures. On the other hand, *Gi* for the PB/PS blends rather strongly depends on the vinyl content of the polybutadiene chains, which, of course, is due to the corresponding entropic Γ_σ contributions (see Eq. 20). The largest *Gi* appear for the blend with dPB(1,2), which is constant up to 100 MP, followed by a rather strong decrease. In the dPB(1,4) blend, pressure induces a linear decrease of *Gi*. The lowest *Gi*

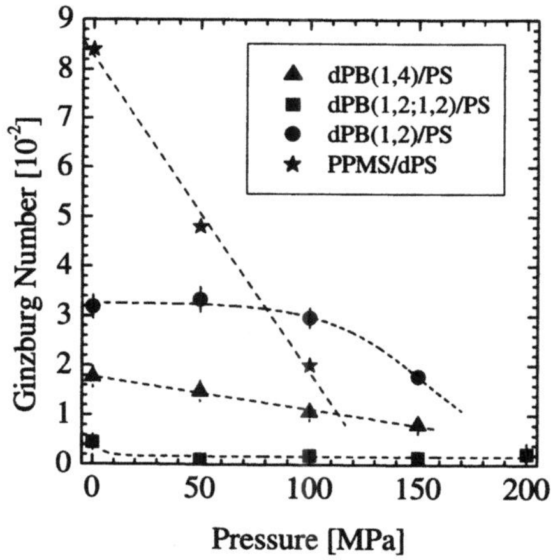

Fig. 14 Ginzburg number versus pressure of the three dPB/PS and PPMS/dPS blends. In all cases pressure reduces the Ginzburg number and thereby reduces the range of strong thermal fluctuations around T_C

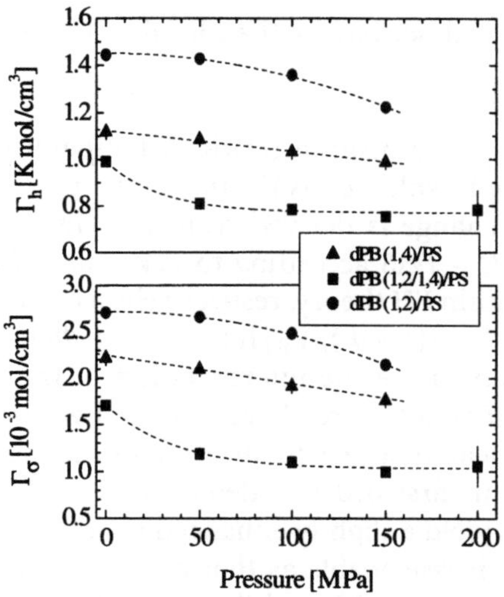

Fig. 15 Enthalpic and entropic parts of FH parameters for the three dPB/PS blends versus pressure. In all cases a decrease with pressure is found. The lowest numbers are found for the PB copolymer mixtures which saturates above 100 MPa

is observed for the copolymer blend for which $Gi = (1.7 \pm 0.5)10^{-3}$ becomes constant above ambient pressure. The last number appears very near the estimated Gi for incompressible polymer blends (see Fig. 8).

It is found from the analysis that the effect of pressure on thermal fluctuations is mainly determined by the variation of the mean field parameters; at least for PB/PS the Ising critical amplitude C_+ (defined above in Eq. 17) is independent of pressure within an accuracy of 3% (Fig. 9 in ref. [37]).

The FH parameters Γ_h and Γ_σ derived from the mean field critical amplitude C_{MF} in Eq. 11 are depicted in Fig. 15. In all cases Γ_h and Γ_σ decrease with pressure. The enthalpic term is largest for the PB(1,2)/PS blend. This can be understood from the solubility parameters of the monomers; the PS chains have a larger solubility parameter than both PB monomers as can be seen from Table 5 in ref. [77]), while the solubility parameter of PB(1,2) is smaller than PB(1,4) as discussed in context with Fig. 11. Therefore, mixing PB(1,2) and PS leads to a larger difference in solubility parameters and thereby to a larger enthalpic term. The Γ_h of the copolymer PB(1,2;1,4) is lowest. But the low entropic term counters a stronger enhanced compatibility from Γ_h.

4.4.2
Clausius–Clapeyron Equation

Within a mean field approximation the pressure induced changes of the binodal and critical temperature are determined by the corresponding changes of the FH parameter as described by the Clausius–Clapeyron equation.

$$\partial_P T_C = T_C \Delta V / \Delta H \tag{25}$$

The change of the phase transition (critical) temperature T_C with pressure depends on the sample volume (ΔV) and enthalpy (ΔH) of mixing [78]. The sample volume change is derived from the Gibbs free energy of mixing $\Delta G(T, P, \Phi) = \Delta H - T\Delta S$ according to $\Delta V = \partial_P \Delta G$ and to the enthalpy and entropy terms within FH theory, respectively, as $\Delta H/R = \Phi(1-\Phi)\chi_h$ and $\Delta S/R = -\{(\Phi/V_A)\ln\Phi + [(1-\Phi)/V_B]\ln(1-\Phi)\} + \Phi(1-\Phi)\chi_\sigma$.

At the critical temperature the interpretation of the thermodynamic parameters ΔV and ΔH is not correctly represented within FH theory because thermal fluctuations lead to a renormalization of the mean field critical temperature T_C^{MF} which in first order is determined as $T_C = T_C^{MF}(1 - Gi)$ [10]. Thermal fluctuations yield a slightly enhanced miscibility whose degree is determined by *Gi*. This is reasonable as thermal fluctuations also enhance the entropy. At elevated pressure this stabilization effect is reduced as is visible from *Gi* in Fig. 14.

Also including contributions from thermal composition fluctuations the Clausius–Clapeyron equation for critical polymer blends can be expressed as [40]

$$\partial_P T_C = T_C(1)\left[\partial_P \Gamma_h - T_C(1)\partial_P \Gamma_\sigma\right]/\Gamma_h(1) - \left[T_C(1)/(1 - Gi(1))\right]\partial_P Gi \tag{26}$$

The first term is determined from the effective FH parameter Γ, while the second term describes the stabilization effect of the thermal fluctuations. The parameters $T_C(1)$, $Gi(1)$, $\Gamma(1)$ represent the corresponding values at the ambient pressure field of 0.1 MPa.

Figure 16 compares $\partial_P T_C$ as experimentally determined from the data in Fig. 13 and as calculated with Eq. 26 from the Ginzburg numbers and FH parameters in Figs. 14 and 15 for all vinyl contents. The full dots show $\Delta_P T_C$ ($\equiv \Delta T_C/\Delta P$) directly derived from the phase boundaries while the open symbols represent the evaluated numbers. The two triangular symbols show the changes caused separately from the FH parameter and the Ginzburg number, while the open dots represent the sum of both effects. For the two samples with smaller vinyl contents the effect from the FH parameter is about twice as large as from the Ginzburg number *Gi*, while for the dPB(1,2)/PS sample a negative $\Delta_P T_C$ is obtained from $\Gamma(P)$ which is more than compensated by the $Gi(P)$ contributions, giving an overall positive $\Delta_P T_C$. A comparison of the "total" $\Delta_P T_C$ from the Clausius–Clapeyron equation with the experimental one shows good agreement.

The effect of pressure on thermal fluctuations in PB/PS always seems to lead to a unique behavior, namely, to an increase of the critical temperature. But the situation is more complex if one considers the contribution from the pressure dependence of the FH parameter $\Gamma(P)$ alone and as already evident from the application of the Clausius–Clapeyron equation. The enthalpic term

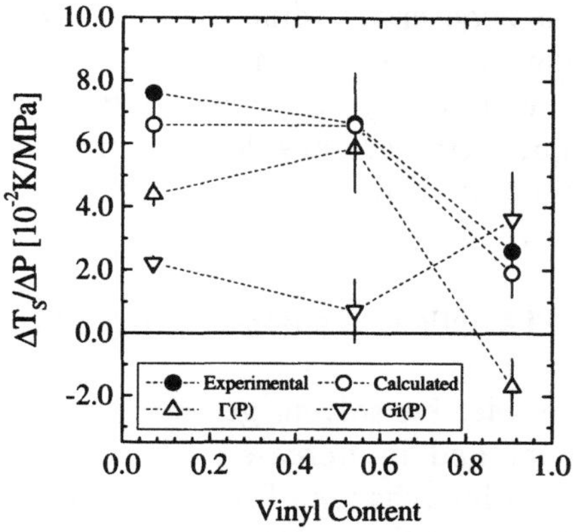

Fig. 16 Pressure induced changes of the spinodal temperature as obtained from experiment (Fig. 13) and from an extended Clausius-Clapeyron equation with the enthalpic and entropic terms in Fig. 15 and the Ginzburg numbers in Fig. 14. Both values are depicted as *solid and open points* which show good agreement and demonstrate consistency

is always positive, and both FH parameter terms Γ_h and Γ_σ decrease with pressure. Therefore, one expects at low temperature a negative change of the critical temperature with pressure when Γ_h becomes the leading term. From the data in Fig. 15 one evaluates the characteristic temperatures of 14 °C, 60 °C, and 132 °C for the dPB(1,4), dPB(1,2;1,4), and dPB(1,2) samples, respectively, below which a negative $\Delta_P T_B$ is expected. So, because the first two samples have critical temperatures larger than 14 and 60 °C, a positive shift of $\Delta_P T_C$ is evaluated in agreement with experimental findings. For the dPB(1,2) sample, however, the critical temperature of 101 °C is smaller than the calculated 132 °C, and therefore a negative change of the critical temperature with pressure is predicted from the contribution of $\Gamma(P)$ alone (see discussion of Fig. 16).

The good agreement of $\Delta_P T_C$ from experiment and the Clausius-Clapeyron equation represents a consistency check. This finding shows that the underlying theory of the crossover function in Eqs. 15–19 and the subsequent analyses lead to reasonable values of the FH parameter and Ginzburg number. Furthermore, it is a clear demonstration that pressure induced changes of the phase boundaries are generally influenced by two pressure dependences, namely, those of the interaction parameter and the degree of thermal composition fluctuations.

While the pressure induced increase of the two-phase boundary seems to be the normal case, there exist a few polymer blends which show an enhanced miscibility with pressure. Such a behavior was observed in PEP/PDMS (poly(ethylenepropylene)/polydimethylsiloxane) and PDMS/PEMS (polyethyl-methylsiloxane) blends [42, 43]. In both blends an increase of the enthalpic and entropic terms of the FH parameter with pressure, i.e., to

$\partial_P \Gamma_h > 0$ and $\partial_P \Gamma_\sigma > 0$, is observed while the total FH parameter is decreasing ($\partial_P \Gamma < 0$) as the entropic term is dominant at the transition temperature. For those blends one expects a positive $\Delta_P T_C$ at sufficiently low temperature when the enthalpic term becomes dominating. The microscopic origin of such behavior is still unknown.

4.5 Binary Blends with Small Additions of a Non-Selective Solvent

The addition of a non-selective solvent generally leads to a larger compatibility of the binary blend and an increase of the Ginzburg number. Pressure experiments visualize an interchange of solvent molecules with free volume. After a short presentation of the scattering law, a series of experiments is discussed.

4.5.1 Structure Factor

The scattering cross-section of a binary polymer blend with small addition of a non-selective solvent is given as

$$d\Sigma/d\Omega(Q) = K_{AB}\Phi_P S(Q) + K_{PS} S_{PP}(Q) \tag{27}$$

The first term is determined by the polymer structure factor given in Eq. 9 and is proportional to the contrast factor K_{AB} of the two polymers and the total polymer concentration Φ_P. The second term contains the polymer-polymer partial structure factor $S_{PP}(Q)$ describing the overall polymer density fluctuations which would become zero for an incompressible polymer melt with $\Phi_P = 1$. The scattering contrasts $K_{AB} = (\rho_A - \rho_B)^2/N_A$ and $K_{PS} = (\bar{\rho} - \rho_S)^2/N_A$ are determined by the coherent scattering length densities of the polymers and solvent molecules, the average $\bar{\rho} = (\Phi_A\rho_A + \Phi_B\rho_B)/\Phi_P$, and Avogadro's number N_A (see given values in Table 1). As we are interested in investigating thermal composition fluctuations, $S(Q)$ has to become the dominating scattering contribution. The relevant contrast K_{AB} can be made an order of magnitude larger than K_{PS} by, e.g., mixing deuterated PB with protonated PS and solvent molecule.

Within the FH model the susceptibility and correlation length are, respectively, given as $S^{-1}(0) = 2(\Gamma_C - \Phi_P\Gamma)$ and $\xi^{-2} = [2(\Gamma_C - \Phi_P\Gamma)]/R_g^2\Gamma_C$. In the "dilution approximation" of blend–solvent systems, Γ is replaced by $\Phi_P\Gamma$ [79]. Such a replacement delivers the mean field critical amplitudes and Ginzburg criterion in terms of the FH parameters according to

$$C_+^{MF} = 1/(2\,|\Gamma_C + \Phi_P\Gamma_\sigma|) = T_C^{MF}/2(\Phi_P\Gamma_h) \tag{28}$$

$$\xi_0^{MF} = \sqrt{2R_g^2\Gamma_C C_{MF}/3} = R_g/\sqrt{3(1 + \Phi_P\Gamma_\sigma/\Gamma_C)} \tag{29}$$

and

$$Gi \propto \left[V^{(2-\gamma)}(2/V + \Phi_{\mathrm{P}}\Gamma_{\sigma})\right]^{1/(\gamma-1)} \propto \left[V^{(2-2\nu)}(2/V + \Phi_{\mathrm{P}}\Gamma_{\sigma})\right]^{1/(2\nu-1)} \quad (30)$$

4.5.2
Susceptibility and Correlation Length

The susceptibility and correlation length of several dPB(1,4)/PS blends (sample 2 in Table 2) mixed with the non-selective solvent ortho-dichloro-benzene (oDCB) between zero and 16% volume fraction are depicted in Fig. 17 versus inverse temperature. At the maximum solvent content the polymer miscibility has been appreciably improved as evident by the strong decrease of the critical temperature from about 70 to 30 °C [32]. The susceptibility and correlation length show the typical deviation from linear mean field behavior, and the results of the fit with the crossover functions of Eqs. 19 and 16 are shown by the solid lines; the agreement between experiment and theory is always good. In nearly all samples the critical condition can well be approached.

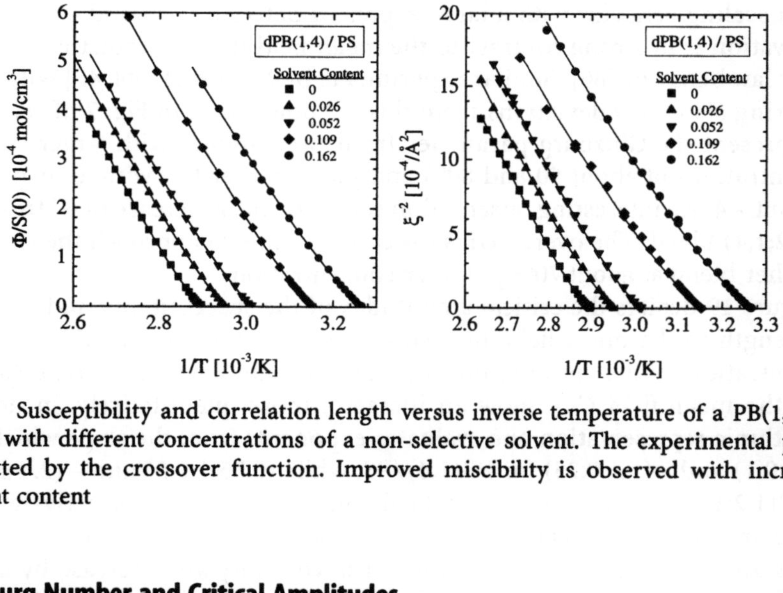

Fig. 17 Susceptibility and correlation length versus inverse temperature of a PB(1,4)/PS blend with different concentrations of a non-selective solvent. The experimental points are fitted by the crossover function. Improved miscibility is observed with increased solvent content

4.5.3
Ginzburg Number and Critical Amplitudes

The effect of solvent on the Ginzburg number is shown in Fig. 18 for the two dPB/PS blends that differ in the vinyl content of the PB component (Sample 2 and 3 in Table 2). *Gi* is plotted versus the total polymer concentration on a double logarithmic scale. A strong increase of *Gi* with solvent content is observed; at 80% polymer concentration a 5 to 10 times larger *Gi* is observed and over this concentration range the Ginzburg number approximately fol-

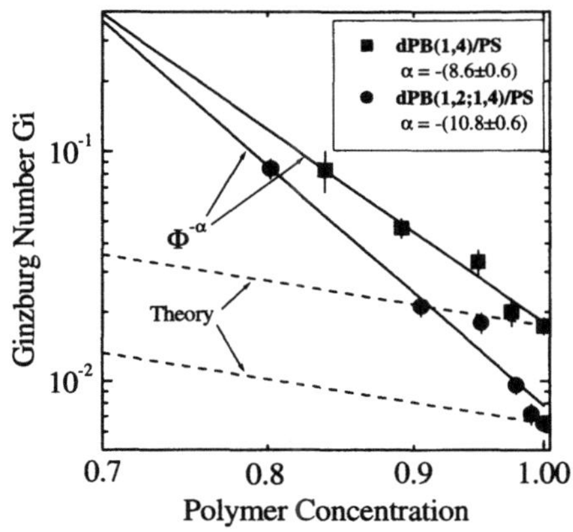

Fig. 18 Ginzburg number versus total polymer concentration in double logarithmic representation. As indicated by the *dashed lines* a much stronger increase of *Gi* with solvent content is observed and is described by an exponent between 8 and 11 instead of 2 as is predicted from theory

lows a scaling behavior with a large exponent between – 8.6 and – 10.8. This observation is in strong contrast to the mean field theory prediction of a – 2 power law behavior [80]. Similar experiments by Nose et al. [44, 45] with light scattering also show deviations from this – 2 power law; in Fig. 7 of ref. [44], they present the Ginzburg parameter from two samples with larger solvent concentrations of about 50 and 80% and find a power law with an exponent of about – 4. An interesting observation is the stronger increase of *Gi* from the PB(1,2;1,4) blend. The overall *Gi* is lower and seems to approach the *Gi* from the other blend at about 70% polymer concentration.

Figure 19 depicts the critical amplitudes of the susceptibility and correlation length in the mean field and Ising approximations versus the polymer concentration. While the amplitudes C_+ show a slight linear decrease by about 20%, the mean field C_{MF} show an increase; more quantitatively, in double logarithmic representation one evaluates exponents from the slope which are (1.1 ± 0.1) and $-(1 \pm 0.25)$ for PB(1,4)/PS, while (0.35 ± 0.5) and $-(2.4 \pm 0.5)$ for PB(1,2;1,4)/PS within the mean field and Ising regimes, respectively.

The mean field critical amplitudes of the correlation length ξ of both blends are the same within experimental uncertainty and increase by about 10% for a 80% polymer content. The "Ising" amplitudes are smaller than their mean field numbers, and, instead, they decrease with polymer content by about 20%. In addition, the amplitudes are different in both blends; PB(1,4)/PS shows slightly more than 10% larger values than PB(1,2;1,4)/PS. The mean field amplitude ξ_0 is proportional to the polymer radius of gyration and the square root of $(1 + \Phi_P \Gamma_\sigma / \Gamma_C)$ (Eq. 29); the polymer parameters are very similar, (Table 2) so the result of equal amplitudes is not surprising. On the other hand, the Ising critical amplitudes seem to depend more sensitively

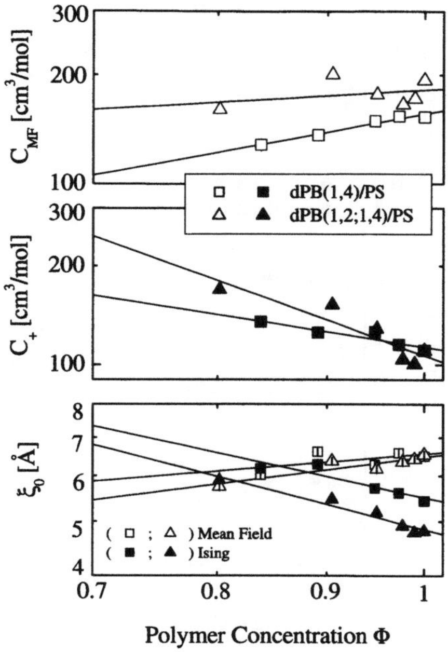

Fig. 19 Critical amplitudes C of the susceptibility and ξ_0 of the correlation length versus total polymer concentration. The "Ising" C slightly decreases by about 10%, while the mean field amplitude shows a slightly stronger relative increase of 20%. The critical amplitudes of the correlation length in the lower figure exhibits an increase of the mean field amplitudes and a decrease of the "Ising" amplitudes with polymer concentration

on the microstructure of the PB polymer component and may be affected by the smaller segment length of the PB with higher vinyl content; a 62% vinyl content PB shows a 9.3% smaller segment length than a 7% one [38].

In Fig. 20 the FH parameters of the blend-solvent systems have been depicted versus the solvent content. In the upper figure the enthalpic Γ_h of both blends are presented for ambient pressure, while in the two lower figures the entropic Γ_σ is depicted at ambient and 200 MPa pressure fields. At ambient conditions one observes for the PB(1,4) blend a linear increase of about 50% for Γ_h and $\Gamma_\sigma$while for the (1,2;1,4) copolymer blends yield an oscillatory behavior with relative maximum and minimum values at about 3 and 12% solvent concentration, respectively. The difference of Γ_σ at the two pressure fields may give information about the free volume. One can reasonably assume that practically no free volume is "surviving" at 200 MPa. We proceed with this analysis by applying a simplified ansatz $\Gamma_\sigma = \Gamma_\sigma^0 + \Phi_S/\Omega_S + \Phi_V/\Omega_V$ with Ω_i and Φ_i, respectively, the molar volume and volume fraction of the solvent molecules and the units of free volume [36]. The straight line for $\Gamma_\sigma$at 200 MPa describes the function $\Gamma_\sigma = \Gamma_\sigma^0 + \Phi_S/\Omega_S$ as the contribution from the solvent molecules, while its difference from the ambient pressure value

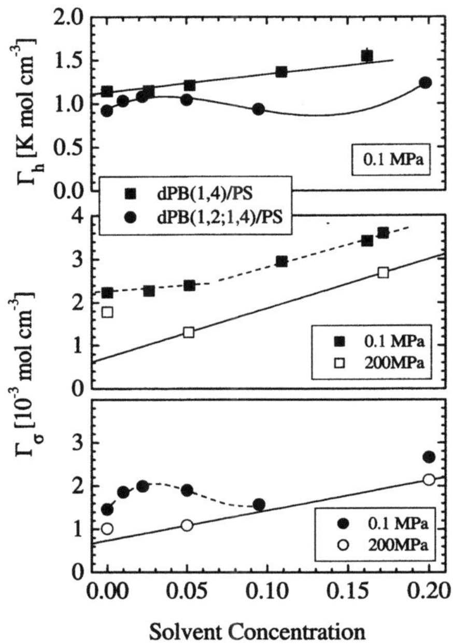

Fig. 20 FH parameter of two PB/PS blends (same as in Fig. 19) versus solvent concentration and at ambient and 200 MPa pressure fields. The Γ_σ at the two pressure fields can be used to estimate the volume fraction of cavities and their compensation by the non-selective solvent molecules

is described by the Φ_V/Ω_V term. According to this ansatz a 10 to 15% free volume content is evaluated for the solvent free sample, and this content continuously decreases with solvent content. Above 10% solvent content no free volume was observed for the copolymer blend, while a constant fraction of about 8% is found in the PB(1,4) sample. These experiments clearly show an interrelation between free volume and non-selective solvent.

5 Crossover to the Renormalized 3D-Ising Critical Behavior

In some polymer blends an additional crossover from 3D-Ising to the so-called renormalized 3D-Ising behavior is observed near the critical point. Such a crossover is depicted in the upper and lower Fig. 21 for the susceptibility and correlation length of two PB/PS blends (Samples 2 and 4 in Table 2) [32, 81]. The characteristics of such a crossover are larger critical exponents in comparison to the Ising case and a shift of the critical temperature. This type of crossover behavior has been systematically studied by the group of Nose in polymer blend-solvent systems [45]. In those systems the

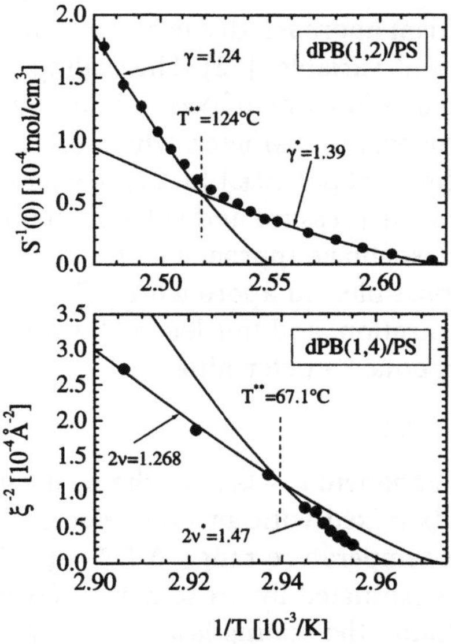

Fig. 21 Crossover of susceptibility and correlation length from Ising to the renormalized Ising behavior near the critical point as observed in the dPB(1,2)/PS and dPB(1,4)/PS blends. The crossover is much more pronounced in the PB(1,2) blend than in the PB(1,4) blend. A selectivity of the cavities in the PB(1,2) blend might be an explanation of the larger effect

renormalized Ising behavior is interpreted based on the appearance of the total polymer density fluctuations near the critical point. So, in order to interpret the observations of the renormalized Ising behavior in "pure" polymer blends, one might think of the stronger polymer density fluctuations and the cavities or excess free volume distribution as a third component.

5.1
Hidden Variables – Fisher Renormalization

Binary polymer blend–solvent or blend-cavity samples can be considered as quasi binary mixtures as long as the total polymer concentration $\Phi_P = \Phi_A + \Phi_B$ is constant and thermal fluctuations are only considered for the order parameter represented by the composition of one component. Under such conditions, the system obeys Ising critical behavior asymptotically close to the critical temperature T_C as discussed in the former sections. If the Gibbs free energy of mixing ΔG is considered as the relevant thermodynamic potential, the equilibrium state is determined by the order parameter Φ, the temperature and the pressure fields. In polymer-solvent systems one has in addition the solvent volume fraction Φ_S and its conjugate chemical potential $\mu_S = \partial_{\Phi_S} \Delta G|_{T,P,\Phi}$ (polymer and solvent concentrations give $\Phi_P + \Phi_S = 1$). In the so-called isomorphism approach discussed by Anisimov et al. μ_S and Φ_S are considered as "hidden" variables [82]. The scaling form of the ΔG "fluctuation" part is described as a function of two relevant parameter fields, namely, the temperature and the order parameter, which are both a function of pressure and the solvent chemical potential [9, 82]. Generally, one observes Ising critical exponents when the pressure and solvent chemical potential are constant during a so-called isochoric approach to T_C. Such an approach, however, is experimentally not possible. An approach to T_C can only be realized with constant solvent concentration, and this leads to a further crossover with the renormalized critical exponents determined as

$$\gamma^* = \gamma/(1-\alpha) \quad \nu^* = \nu/(1-\alpha) \tag{31}$$

In terms of the critical exponent $\alpha = 0.11$ of the heat capacity [83].

For polymeric blends mixed with an non-selective solvent the transition temperature T^{**} or the temperature range $\Delta T^{**} = (T^{**} - T_C)$ to a renormalized Ising behavior was estimated by Broseta et al. [80]; on the basis of mean field parameters they found the scaling law $\Delta T^{**} \cong N^{-13.6}\Phi^{-17}$. This expression leads to an extremely small temperature interval around T_C, which is not visible in our experiments, and suggests the behavior as being that of a quasi binary blend characterized by Ising universality near T_C.

5.2 SANS Results on Blends

The blend-solvent systems, whose susceptibility and correlation length are depicted in Fig. 17, show a crossover from mean field to Ising critical behavior and are well described by the crossover functions (Eqs. 15 to 19), and no further transition to Fisher's renormalized Ising behavior is observed. This means that all blend-solvent samples can be considered as quasi-binary blend systems in accord with the former discussion. The observation of a crossover to a Fisher renormalized Ising behavior in the two solvent free dPB/PS blends as depicted in Fig. 21 is therefore surprising. The susceptibility of the dPB(1,2)/PS blend is plotted showing the crossover at $T^{**} = 124\,°C$ with the critical exponents $\gamma^* = 1.39$ and $2\nu^* = 1.43$ in agreement with the theoretically expected number. In this blend the crossover occurs over a large temperature range of about 18 K and leads to an about 12 K larger range of miscibility. The crossover observed for the correlation length of the dPB(1,4)/PS blend as depicted in the lower figure looks qualitatively different and a slightly larger exponent than expected from theory was determined. The crossover starts at about 2 K above T_C and leads to a slight enhancement of the critical temperature as indicated by the dotted line. It has to be mentioned that in order to observe a renormalized crossover, the experiments have to be performed very slowly over several hours in small temperature steps; so in a second and faster experimental run no renormalized crossover was observed for the same

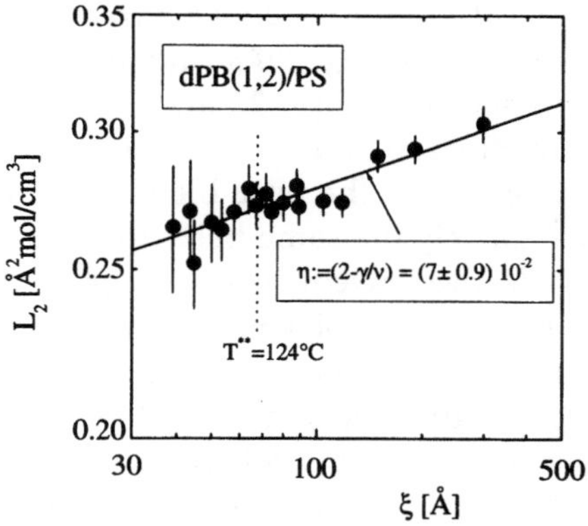

Fig. 22 The coefficient L_2 of the Q^2 term of the structure factor versus correlation length in double logarithmic presentation. According to theory the slope gives the Fisher exponent $\eta = 0.034$ which is found a factor of two larger than predicted. The crossover transition at T^{**} has no effect on η as the critical exponents γ and ν are renormalized in a same manner

dPB(1,2)/PS sample, and the phase transition occurred at a critical temperature as extrapolated from the $\gamma = 1.24$ scaling law. This means that density fluctuations in polymer melts must have large relaxation times of the order of hours.

An additional proof of the crossover from Ising to renormalized Ising critical behavior might be the functional dependence of the coefficient L_2 of the Q^2 term in Eq. 12 versus the correlation length ξ. L_2 is proportional to the chain statistical segment length and follows a scaling law according to $\xi^{\eta=(2-\gamma/\nu)}$ in both regimes of Ising and renormalized Ising universality classes [24]. In Fig. 22 L_2 of the d-PB(1,2)/PS sample has been depicted versus the correlation length in double logarithmic presentation. The data follow quite well a single line with a slope of $\eta = (0.07 \pm 0.01)$ and no transition is observed at the crossover temperature T^{**} which is indicated by the dashed line. The Fisher exponent is about twice as large as the predicted one.

5.3
Renormalized 3D-Ising Critical Behavior in Blends and Blend-Solvent Systems

Crossover to the renormalized Ising critical behavior was observed in two symmetric ($\Phi_A \cong \Phi_B$) PB/PS polymer blends without any solvent but not in the corresponding blends mixed with the non-selective solvent oDCP up to a concentration of 20 vol %. It might be instructive to compare the observations in both blends with results from polymer blend-solution mixtures obtained by the group of Nose [44,45]. In their experimental work no crossover was found in symmetric blend-solvent samples in accord with our SANS results. A transition to the renormalized Ising case was observed in two asymmetric blends with $\Phi_A \neq \Phi_B$ and mixed with a selective and non-selective solvent as shown in Figs. 22 and 23 of ref. [44]. Both samples show a characteristically different crossover behavior which looks very much like those observed in Fig. 21 for the pure blends. The blend with the selective solvent shows qualitatively the same behavior as dPB(1,2)/PS, while the crossover of the non-selective solvent-blend sample is small and similar to that in the PB(1,4)/PS sample. This observation seems plausible as a selective solvent is stronger correlated with polymer composition fluctuations. From these findings one might argue that cavities or free volume represent a third component in polymer blend systems, which in some cases may lead to the observed renormalized critical behavior. Furthermore, the observation of different renormalized crossover characteristics in the upper and lower parts of Fig. 21 might give a hint of a distinct selectivity of the cavities with respect to the PB(1,4) and PB(1,2) monomers. So, the absence of a renormalized Ising crossover in the PB/PS/solvent samples might be understood from a compensation of the cavities by the non-selective solvent as discussed in context with the pressure experiments in Fig. 20.

6
Crossover to Isotropic Lifshitz Critical Behavior in (A/B) Polymer Blend/(A-B) Diblock Copolymer Mixtures

Mixing two partially incompatible (A/B) homopolymers with small amounts of an (A-B) diblock copolymer leads to an improved miscibility in a similar way as amphipathic molecules do in oil-water mixtures [for convenience we will abbreviate those systems as (A;B)]. At larger diblock content, microemulsion and ordered phases are formed. A reason of this rather complex phase behavior is that the two constituents belong to different universality classes. Diblock copolymers are of Brasovskii type [53]; they decompose into spatially modulated phases of mesoscopic length scale determined by the polymer size, while homopolymer blends are of 3D-Ising type and decompose into macroscopically large domains (see discussions in context with Fig. 1). Within mean field approximation these two phases meet at the Lifshitz line and the corresponding critical point. An important parameter is the ratio of the homopolymer and copolymer molar volumes $\alpha = \sqrt{V_A V_B}/V_{A\text{-}B}$. We will limit our discussions to systems with $\alpha \approx 0.17$; in those systems one expects no tricritical Lifshitz point and no three-phase liquid region [55].

A first systematic study of such systems with SANS was performed on the relatively large molar mass symmetric polyolefins PE and PEP and the corresponding diblock copolymer PE-PEP, PE being polyethylene and PEP being poly(ethylene propylene) [46]. A mean field Lifshitz-like behavior was observed near the predicted isotropic Lifshitz critical point with the critical exponents $\gamma = 1$ and $\nu = 0.25$ of the susceptibility and correlation length, and the structure factor following the characteristic mean field Lifshitz behavior according to $S(Q) \propto Q^{-4}$. Thermal composition fluctuations seem not to be relevant here from the observation of the mean field critical exponents. On the other hand, fluctuations seem to exercise an influence as the Lifshitz critical point is not observed and instead a one-phase channel of a polymeric bicontinuous microemulsion phase appears [46], which could be clearly identified by transmission electron microscopy (TEM) [47]. Equivalent one-phase "channels" were also observed in other systems [48–50]. In a quite recent study on (dPB;PS), which will be discussed in more detail below, the microemulsion channel is separated be another Lifshitz line into a droplet and bicontinuous microemulsion phases [52, 84].

As this article is devoted to critical crossover phenomena in polymer blends, we limit our considerations to homopolymer blends of critical composition mixed with diblock concentrations below the Lifshitz line.

6.1
Phase Diagram of a (A/B) Polymer Blend/(A-B) Diblock Copolymer Mixture

The temperature-copolymer concentration plane at fixed ambient pressure of a (dPB;PS) blend phase diagram (sample 11 in Table 2) below $\Phi \cong 0.1$ diblock content is depicted in Fig. 23. The homopolymer dPB/PS blend of critical composition was mixed with a symmetric dPB-PS diblock copolymer of about six times larger molar volume giving $\alpha = \sqrt{V_A V_B}/V_{A\text{-}B} \cong 0.16$. This phase diagram is quite typical for those systems. It is divided into several sections. At high temperature one has two homogeneous (disordered) phases which are separated by the Lifshitz line (LL). At concentrations below the LL at $\Phi < \Phi_{LL}$ the structure factor $S(Q)$ looks similar to that of binary blends with its maximum at $Q = 0$. Above the Lifshitz line $\Phi > \Phi_{LL}$ one observes the characteristics of diblock copolymer melts with the maximum $S(Q^*)$ at a finite $Q = Q^*$. In all cases $S(Q^*)$ represents a susceptibility with $Q^* = 0$ and $Q^* \neq 0$ below and above Φ_{LL}, respectively. Within mean field approximation the Lifshitz line is estimated at a constant concentration according to $\Phi_{LL} = 2\alpha^2/(1 + 2\alpha^2)$ and should here be observed at $\Phi_{LL} = 0.048$. The experimental Lifshitz line, however, shows characteristic deviations; it is found at slightly larger diblock concentrations and shows a pronounced dependence on temperature near the two-phase boundary. At high temperatures the Lifshitz line approaches a constant value of about 0.06; at around 70 °C it displays its largest value of 0.084, and approaches a value of 0.07 at low temperatures. A similar observation has been reported in ref. [48, 49] and will later be explained as due to the effects of thermal composition fluctuations [57].

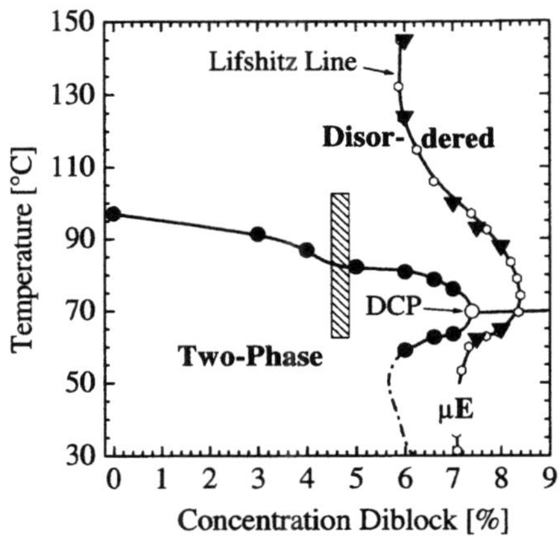

Fig. 23 Phase diagram in the temperature-diblock copolymer plane for the (dPB;PS) mixture below the Lifshitz line separating blend like from diblock-like phase behavior. The *full dots* and the *solid line* represent the critical points of a two-phase region. The *hatched area* indicates a crossover from Ising to isotropic Lifshitz critical behavior, and a double critical point DCP is at 7% diblock concentration. The Lifshitz line separates at high and low temperatures the disordered phases and droplet and bicontinuous microemulsion phases (μE). Its non-monotonic shape near the DCP is caused by the strong thermal fluctuations

At diblock concentrations above Φ_{LL} structures of bicontinuous microemulsion (μE) and of lamella ordered phases were identified [46]. While the lamella phase was predicted by mean field theory, the observation of a microemulsion phase was unexpected. Polymer microemulsion phases are presently the subject of active research both from theory [85] and experiment [52] and, in addition, seem to generate interest for industrial application because of its homogeneous structure on the μm length scale.

Below the Lifshitz line ($\Phi < \Phi_{LL}$), a line of critical points has been determined (full dots in Fig. 23) separating the homogeneous from the two-phase region. This phase boundary of critical temperatures was determined from the SANS susceptibility in a similar way to that described in the context of Fig. 5. These blends always fulfill the conditions of criticality as binodal and spinodal curves were observed at the same temperature. The solid line in the phase diagram is a guide for the eye and represents the critical points which decrease with increasing diblock concentration and terminate at a co-called double critical point (DCP) at about $\Phi = 0.07$. In the range between $\Phi = 0.06$ and 0.07, both an upper and lower critical solution phase boundary is observed. Below the LCST boundary a droplet microemulsion phase is formed which at higher diblock content above LL transforms into a bicontinuous microemulsion phase.

Two critical universality classes are observed along the line of critical points and are separated by the dashed area at $\Phi = 0.048$. Below $\Phi < 0.048$

a mean field to Ising crossover is present, while above $\Phi > 0.048$ a mean field to isotropic Lifshitz critical behavior crossover is observed. Finally, above $\Phi > 0.06$ the effect of the double critical point becomes visible. Quite generally, the Lifshitz critical point is a multiple critical point connecting the phase boundaries of macro-phase separation and of disorder-order transition. Such a critical point, however, is only realized within a mean field approximation. Because thermal fluctuations are relevant, such a Lifshitz critical point cannot exist. This is because of the different strength of fluctuations above and below Φ_{LL}, expressed by the Ginzburg criteria for blends in Eq. 20 and for diblock copolymers in ref. [56].

6.2 Structure Factor within Mean Field Approximation

Here we have the situation that one of the (A/B) monomers of the homopolymer and of the copolymer was always deuterated with the same relative amount of deuterium. Under such conditions the structure factor $S(Q)$ measures thermal composition fluctuations with respect to the total monomer fraction, which corresponds to a scalar order parameter represented by the local concentration $\Phi = \Phi(x)$ of the A or B monomers. The basic thermodynamic features of those systems near their binodal line are well described by the common Landau expansion of the free energy according to

$$H = \int \mathrm{d}^d x \left\{c_2(\nabla\Phi)^2 + c_4(\nabla^2\Phi)^2 + r\Phi^2 + u\Phi^4 + u_6\Phi^6\right\} \tag{32}$$

representing a functional of the order parameter [9, 54, 55]. This Hamiltonian H represents an extended form of Eq. 14 that includes higher expansion parameters in the gradient energy and in the order parameter. A principal effect of the dissolved diblock copolymers is the reduction of the surface energy which according to the Hamiltonian is described by a smaller parameter c_2. This parameter is positive at low copolymer concentration, becomes zero at the Lifshitz line Φ_{LL}, and is negative for larger copolymer contents.

The Hamiltonian accounts for thermal composition fluctuations in the disordered regime. For positive c_2-values the structure factor $S(Q)$ has the characteristic behavior of polymer blends: $S(Q)$ is maximum at $Q = 0$, and the susceptibility r^{-1} is correspondingly given by this $S(Q = 0)$ value: $r^{-1} = 2S(0)$. For negative c_2-values the structure factor $S(Q)$ has the basic characteristics of block copolymer melts, i.e., the maximum value of $S(Q)$ appears at a finite Q-value, $Q = Q^*$. The susceptibility is then given by the structure factor at this Q^*-value. Mean field theory for symmetric copolymers predicts that $S(Q^*)$ will diverge at a critical point, and beyond that the system will order on a mesoscopic length scale through microphase separation. The Lifshitz critical point is determined by two conditions, namely $c_2 = 0$ and $S^{-1}(0) = 0$. With $c_2 \cong 0$ in the vicinity of the Lifshitz point, the fourth order term of the gradient energy, c_4, becomes a leading term in the free energy, giving rise to the characteristic Lifshitz $S(Q) \propto Q^{-4}$ behavior of the structure factor. A Lifshitz tricritical point would be realized in the case of the three conditions, namely $S^{-1}(0) = 0$, $c_2 \cong 0$, and $u = 0$ and u_6 and c_4 all positive.

The structure factor for the polymer blend-diblock copolymer system is described within the random phase approximation by [54–56]

$$S^{-1}(Q) = F(Q)/V - 2\Gamma \tag{33}$$

where Γ is the FH parameter and $F(Q)$ the inverse form factor. $F(Q)$ can be calculated in terms of the partial structure factors S_{AA}, S_{BB} and S_{AB} of all A and B monomers and describes the correlation between the A/B monomers,

$$F(Q)/V = \frac{S_{AA}(Q) + S_{BB}(Q) + 2S_{AB}(Q)}{S_{AA}(Q)S_{BB}(Q) - S_{AB}^2(Q)} \tag{34}$$

For the ternary systems discussed in Fig. 23, $F(Q)$ can be simplified to

$$F(Q)/V = 2/[S_{AA}(Q) - S_{AB}(Q)] \tag{35}$$

for A-B diblock copolymers with molar volume V and homopolymers of equal molar volume $V_A = V_B$ and equal partial structure factors $S_{AA} = S_{BB}$. Assuming a polymer conformation of unperturbed Gaussian chains, $F(Q)$ can be written in terms of the Debye-function (Eq. 7) as

$$F(x) = 4/\left[(1-\Phi)\alpha P_{\text{Debye}}(1, x\alpha) - \Phi P_{\text{Debye}}(1, x) + 4\Phi P_{\text{Debye}}(0.5, x)\right] \tag{36}$$

where $x = R_g^2 Q^2$, R_g being the radius of gyration of the diblock copolymer; and α the ratio of the molar volumes defined above. From the minimum of $F(x)$ one gets both the FH parameter Γ_S at the spinodal and critical point, and the corresponding characteristic $Q = Q^*$ value [54–56]. The structure factor in Eq. 33 can be expanded into powers of Q^2 according to

$$S^{-1}(Q) = S^{-1}(0) + L_2 Q^2 + L_4 Q^4 + \ldots \tag{37}$$

with coefficients determined from the parameters of the Hamiltonian (Eq. 32). The first term is equal to the susceptibility, $S^{-1}(0) = 2(\Gamma_S - \Gamma)$ for concentrations less than the Lifshitz value. The coefficients L_2 and L_4 are proportional, respectively, to c_2 and c_4, and can be determined in terms of the polymer parameters R_g, V, α, and the diblock concentration Φ. At the Lifshitz concentration ($c_2 = L_2 = 0$), a characteristic mean field behavior $S^{-1}(Q) \propto Q^4$ appears from this equation. A realization of such a transition from $S(Q) \propto Q^{-2}$ to $\propto Q^{-4}$ is demonstrated in Fig. 24 showing $S(Q)$ from the (PEE;PDMS) blend in a Zimm representation for three copolymer concentrations below the Lifshitz line [49]. The solid lines represent fits of Eq. 37 from which the three parameters, namely, the susceptibility $S(0)$ and the coefficients L_2 and L_4 are obtained. At $\Phi = 4.3\%$, $S(Q)$ is well described by the Ornstein–Zernike approximation with $L_4 = 0$. For larger Φ the Q^4 term in $S(Q)$ becomes stronger

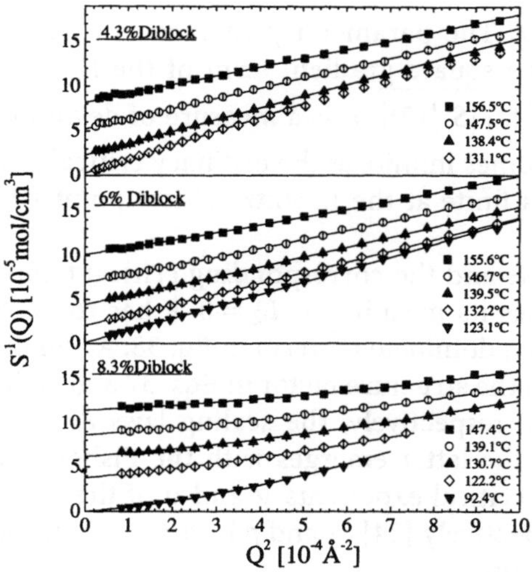

Fig. 24 Structure factor in Zimm representation for the PEE;PDMS blend with different diblock concentrations. The contribution of the Q^2 term decreases with diblock content because of the reduced strength of the surface energy. Instead the next higher expansion term Q^4 becomes relevant and dominates

as demonstrated for the $\Phi = 6\%$ and 8.3% samples. In addition, $S(0)$ is diminished with increasing Φ according to the increased stabilization of the sample miscibility caused by the diblock copolymers.

6.3
Effect of Thermal Composition Fluctuations

Near the Lifshitz line thermal composition fluctuations are expected to become strong over a larger temperature range because of the reduced surface energy ($c_2 \propto L_2 = 0$), leading to a lower threshold force for thermal fluctuations. On a more abstract level this effect can also be interpreted in terms of a larger upper critical dimension $D_U = 8$ beyond which thermal fluctuations become irrelevant, and Gi is twice as large as for ordinary binary polymer blends [86].

Near the critical temperature the structure function can be written in the following form [54]

$$S^{-1}(Q) = S^{-1}(0)[1 + (Q\xi)^2 + Kp^{-2}(Q\xi)^4] \tag{38}$$

with the susceptibility $S(0)$, the correlation length ξ, and the pre-factor Kp^{-2}. The susceptibility follows the well known scaling law $S^{-1}(0) = C_+^{-1}\tau^\gamma$ with the reduced temperature τ and the critical amplitude C_+ (see Eq. 11), the correlation length is determined by $\xi = \sqrt{S(0)L_2}$, and the pre-factor varies as $Kp^{-2} = L_4/(L_2^2 \cdot S(0))$ in terms of the susceptibility and the coefficients L_2 and

L_4 defined in Eq. 37. The parameter p represents a scaling field amplitude determined from the square gradient term of the Hamiltonian (Eq. 32) according to $p = c_2/\sqrt{4c_4 \left|S^{-1}(0)\right|}$. It is a measure of deviation from the Lifshitz critical point; it becomes infinite at the ordinary critical point ($S^{-1}(0) = 0$ and $c_2 > 0$) and remains finite at the Lifshitz critical point when $S^{-1}(0) = 0$ and $c_2 = 0$ [54].

At the Lifshitz line and the corresponding critical temperature, the correlation length ξ looses its meaning as $L_2 = 0$. ξ has to be redefined from the coefficient of the then dominate Q^4 term in Eq. 38. At smaller copolymer contents, the Q^4 term in the structure factor in Eqs. 37 and 38 becomes negligible, and ξ and p follow, respectively, the scaling laws $\xi_0\, t^{-\nu}$and $p^2 \propto \xi^{(2+\eta)}/L_4$. Because of $L_2 \propto \xi^{\eta}$ the latter emerges with the Fisher exponent $\eta = 2 - \gamma/\nu$ evaluated from the critical exponents γ and ν of the susceptibility and correlation length, respectively [34]. ξ and p become accordingly both infinite at the critical temperature.

6.4 SANS Results from below the Lifshitz Line

Figure 25 displays the effects of thermal composition fluctuations on the inverse susceptibility $S(0)$ for a (PEE;PDMS) mixture (sample 10 in Table 2) versus $1/T$ for different diblock concentrations below the Lifshitz line [48]. The critical temperatures determined from $S^{-1}(0) = 0$ decrease with increasing diblock content in a similar way as shown for the (PB;PS) blend (Fig. 23). The $\Phi = 4.3\%$ sample behaves as a pure blend: At high temperatures $S^{-1}(0)$

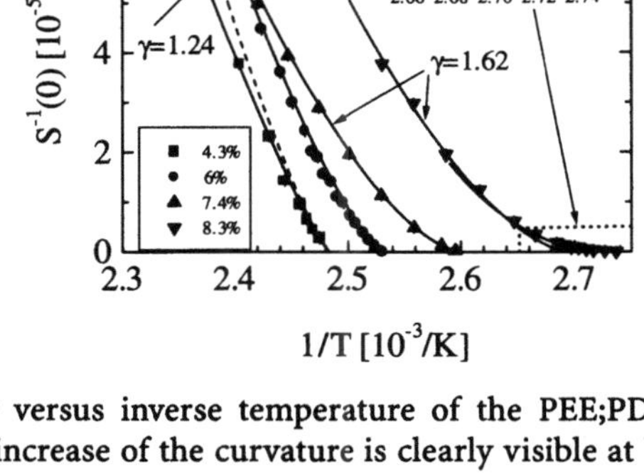

Fig. 25 Susceptibility versus inverse temperature of the PEE;PDMS blend at different diblock content. The increase of the curvature is clearly visible at higher diblock content

shows the mean field and near the critical point the 3D-Ising characteristics with a crossover regime in between as demonstrated by the fitted solid line of the crossover function (Eq. 19). The dashed line in the figure represents the asymptotic 3D-Ising scaling law with the critical exponent $\gamma = 1.24$ as calculated from the fit parameters of the crossover function. The susceptibility of the 0.06 to 0.083 diblock concentration samples shows an increased curvature which can be described by the scaling law $S^{-1}(0) = C^{-1}\tau^{\gamma}$ with the critical exponent γ larger than 1.24 as depicted by the solid lines.

The increase of the γ-exponent indicates a crossover to the universality class of the isotropic Lifshitz case. In contrast to the Ising case of the 4.3% sample (dashed line), an appreciably larger asymptotic critical range is observed as the corresponding asymptotic scaling law describes the susceptibility over the whole experimental temperature range. This observation means a strong increase of the Ginzburg number as qualitatively expected from the corresponding Ginzburg criterion $Gi \propto 1/V^{0.4}$ which shows a much weaker dependence on polymer size than polymer blends ($Gi \propto 1/V$) [54]. A similar analysis was also performed for the correlation length according to $\xi = \xi_0 \tau^{-\nu}$ with the critical exponent ν as published in [49].

In two samples with concentrations 6.7% (not shown in Fig. 25 for clarity) and 8.3% an additional crossover was observed very near the critical temperature in which there is further stabilization of the homogeneous phase, e.g. a lower critical temperature. In this range the susceptibility was analyzed by the same scaling law and is demonstrated for the 8.3% sample in the inset of Fig. 25. Such a crossover was not observed in the 7.4% sample. The latter experiment was performed in larger temperature steps as visualized by the positions of experimental points. Therefore, this sample spent less time in the region of strong thermal fluctuations. This means that two binodal and spinodal lines exist in the range of $0.06 \leq \Phi \leq \Phi_{LL}$; phase separation occurs either at the lower or upper boundary, depending on whether a crossover to the "new" universality class could occur or not. Such a crossover, however, is only observable if the experiments are performed sufficiently slowly. We abbreviate this new universality class as the renormalized Lifshitz critical behavior. An explanation of this crossover might be related to rearrangements of the diblock copolymer as caused by strong thermal composition fluctuations near T_C that would further stabilize the sample against phase decomposition. Such an effect was indeed observed by simulation studies on spinodal decomposition; within the early to intermediate time regime of spinodal phase decomposition the block copolymers became accumulated at the interface of the domains [87]. So, this universality class seems quite analogous to the renormalized 3D-Ising critical behavior as discussed in Sect. 5.

The SANS data from the (PEE;PDMS) mixture between 9 and 11% diblock concentration (not shown here) [49] and the (PB;PS) mixture between 6 and 8% [52] exhibit at low and high temperatures the characteristics of diblock copolymers, i.e. $S(Q)$ shows a maximum at the finite Q^*, while at intermediate temperatures the behavior corresponds to those of homopolymer blends, i.e. $S(Q)$ has a maximum at $Q = 0$. This means that the Lifshitz line was crossed twice as its concentration changes with temperature. This is shown in the phase diagram of the (PB;PS) mixture (Fig. 23); the diblock concentration of the Lifshitz line changes with temperature near the two phase region and is similar to that of the (PEE;PDMS) mixture. The apparent non-monotonic shape of the Lifshitz line could be explained quite recently as due to thermal composition fluctuations. Renormalization group calculations by Kudlay and Stepanow [57] show that at high and low temperatures Φ_{LL} approaches the constant mean field value when thermal fluctuations become negligible, and Φ_{LL} changes with temperature when fluctuations become strong.

In the (PB;PS) mixture a crossover from Ising to isotropic critical Lifshitz behavior was observed at about 4.8% diblock concentration as indicated by the dashed area in the phase diagram of Fig. 23. On the other hand, a crossover to a "renormalized" Lifshitz critical behavior was not observed in this system. The susceptibility critical exponent γ of both systems has been depicted in Fig. 26 versus diblock concentration. A crossover of the exponent γ at about 4.8% and 6.2% from the Ising 1.24 to larger values is visible for the (PB;PS) and (PEE;PDMS), respectively. A constant $\gamma = (1.62 \pm 0.01)$ and

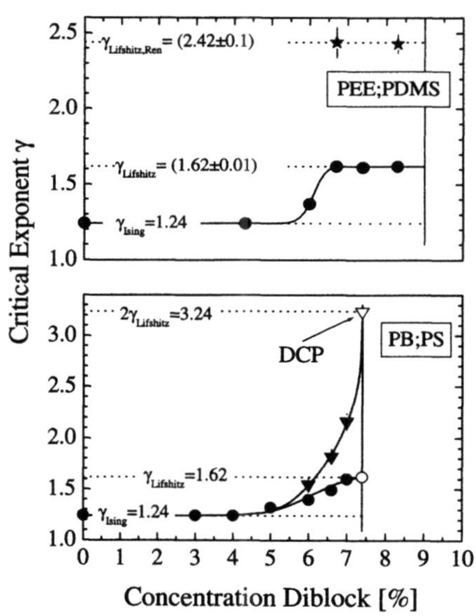

Fig. 26 Critical exponent of the susceptibility γ versus diblock content of two homopolymer blend-diblock copolymer melts. The crossover to the renormalized Lifshitz critical behavior with $\gamma = 1.62$ is clearly visible. Differences of the two blends are remarkable: In PEE;PDMS a plateau of $\gamma = 1.62$ and a further crossover to a "renormalized" Lifshitz behavior is found. In PB;PS, on the other hand a double critical point with the Lifshitz critical exponent $\gamma = 1.62$ is observed

renormalized $\gamma = (2.42 \pm 0.01)$ is obtained for (PEE;PDMS). In (PB;PS) the situation is more complex because of the occurrence of a double critical point at about 7% diblock concentration.

A double critical point (DCP) represents an endpoint of an UCST and LCST line of critical points. The LCST critical line is observed over a limited diblock concentration of about 1%. This means that the homogeneous phases above and below the two-phase regime must represent different phases, as the compatibility is achieved by, respectively, the entropic and enthalpic term of the Gibbs free energy of mixing; the SANS experiments show that the homogeneous phase at lower temperature is represented by a droplet microemulsion phase which must represent a more ordered state than the other two phases at higher temperature.

Double critical points in general and the observed one in the (PB;PS) system are characterized by a strong increase of γ if one applies the scaling law of $S^{-1}(0) = C^{-1}\tau^{\gamma}$ with the "conventional" reduced temperature field. If one, however, applies a modified reduced temperature field according to $\hat{\tau} = (1 - T_{UC}/T)(1 - T_{LC}/T)$ then the inverse susceptibility delivers a critical exponent γ of the corresponding universality class, which here is of the isotropic Lifshitz class [88]. At the DCP one has $T_C = T_{UC} = T_{LC}$ and therefore $S^{-1}(0) = C^{-1}\tau^{2\gamma}$. The solid lines in the lower Fig. 26 describe such a behavior for the isotropic Lifshitz case. The critical exponents shown by the full triangles together with the critical temperatures T_{UC} and T_{LC} given as full points in the phase diagram (Fig. 23) indeed lead to a crossover to the Lifshitz critical exponent $\gamma = 1.62$ a number which was formerly determined for the (PEE;PDMS) system. Up to now the Lifshitz critical exponents in Fig. 26 could exclusively be determined from the SANS experiments discussed in this article. A theoretical determination was not yet possible because the theory lacks a small expansion parameter caused by the large upper critical dimension ($D_U = 8$) [86].

7
Summary and Outlook

In this article we considered thermal composition fluctuations in binary polymer blends under various conditions. Samples of critical composition were studied in temperature and pressure fields, with small additions of a nonselective solvent, or in mixtures with a symmetric diblock copolymer with the same monomers as the homopolymers. Blends of critical composition were chosen in order to follow the fluctuations up to the critical point which represents the stability limit of miscibility. The strength of thermal fluctuations is estimated by the Ginzburg criterion which in the incompressible limit follows the universal scaling law $1/V$ and predicts that binary polymer blends should in principle be described by the FH mean field theory. Recent theoretical considerations, however, show that compressibility or packing effects of the polymer chains lead to much larger Ginzburg numbers and therefore to a stronger effect of thermal composition fluctuations. These considerations are supported by Monte Carlo simulations and scattering experiments.

Most of the data presented here were obtained from small angle neutron scattering (SANS) experiments. The main reason is the strong scattering contrast between the chain components of deuterated and protonated monomers and the low background. The structure factor $S(Q)$ measures the degree of thermal composition fluctuations and delivers the susceptibility $S(Q=0)$ which is linked to the corresponding thermodynamic potential $\Delta G(T, P, \Phi)$ by the fluctuation-dissipation theorem. $S(Q=0)$ follows a scaling law in terms of the reduced temperature in the asymptotic mean field and Ising regimes and becomes infinite at the critical point.

Temperature and pressure fields determine the equilibrium state of a binary polymer blend of fixed composition. Some arguments in the literature state that temperatures and pressures have similar effects on $S(0)$, so that $S(0)$ can be transformed into a universal curve. Such an argument might be true within mean field approximation. However, as the Ginzburg number is non-universal and is a function of pressure, both external fields influence the equilibrium state in different manners. Pressure reduces Gi as the packing of the polymers is increased. In most cases, increased pressure leads to a lower miscibility and to a decrease of the enthalpic and entropic terms of the FH parameter. This means that the temperature of the phase boundary is sufficiently large that the entropic part is dominant. In a few polymer blends, however, an abnormal pressure induced increase of the miscibility is observed. In those blends, in contrast to most of the other blends, an increase of enthalpic and entropic terms is found, again the entropic one the dominate term. The microscopic origin of such opposite behavior of the FH parameter is not yet understood.

Polymer blends mixed with small amounts of a non-selective solvent were explored in both external fields. The Ginzburg number shows a remarkably strong increase with solvent content; a power law of the total polymer content with exponents between − 9 and − 11 is observed. This result is in strong contrast to mean field predictions proposing an exponent of minus two. This again is a strong hint that the Ginzburg criterion in Eq. 20 is strongly affected by the non-combinatorial entropy of mixing Γ_σ, e.g., a larger Γ_σ leads to a huge increase of Gi because of the large exponents in the Ginzburg criterion of Eq. 20. A further observation is that the excess free volume is compensated by the solvent molecules.

The rare case of a crossover to the Fisher renormalized Ising behavior was found in the two PB(1,4)/PS and PB(1,2)/PS blends near T_C (see Figs. 22–23), but these two cases look quite different. Similar observations were made in asymmetric ($\Phi_A \neq \Phi_B$) polymer blend-solvent systems. A large effect was only found if a highly selective solvent was chosen. The dPB(1,4)/PS results shows a crossover very near T_C similarly to their non-selective solvent case, while the crossover in dPB(1,2)/PS qualitatively agrees with the behavior expected when their system is mixed with a highly selective solvent. This coincidence is a hint that cavities of excess free volume, mathematically representing a third component, lead to the observed renormalized critical behavior. One may furthermore speculate about a possible selectivity of voids in the dPB(1,2)/PS blend. No renormalized Ising behavior was observed in the PB/PS/solvent samples which can be understood from compensation of the selectivity of the cavities by the non-selective solvent (see Fig. 20).

A mixture of a symmetrical A-B diblock copolymer and a critical binary A,B homopolymer blend leads to a complex phase diagram representative of the characteristic phase behavior of binary blends and diblock copolymers, namely, phase decomposition into macroscopically large domains and lamella ordered phases on a length scale determined by the diblock size, respectively. This polymer system displays several similarities with mixtures of amphipathic molecules in oil-water. Here we discussed two systems containing a symmetric diblock copolymer with roughly five times larger molar volume than the homopolymers. In these systems an isotropic Lifshitz critical point and Lifshitz line is predicted in the transition range of both ordered phases. An isotropic Lifshitz critical point, however, is not observed as it is destroyed by the different strengths of the fluctuations in homopolymer blends and diblock copolymers. Instead a droplet and bicontinuous microemulsion structure appears in this range. Below the characteristic Lifshitz border line thermal composition fluctuations show a crossover from 3D-Ising to isotropic Lifshitz critical universality. In the (PEE;PDMS) blend a further crossover to a renormalized Lifshitz behavior is observed, while in the (dPB;PS) system a double critical point (DCP) is found.

From these findings it becomes clear that the mean field approximation can only be a poor approximation in those three component polymer mixtures. In particular, critical fluctuations become strong near the isotropic Lifshitz critical point since the stabilizing force from the surface energy becomes small. There are indeed several observations which originate from strong thermal fluctuations such as the microemulsion phase, the destruction of the isotropic Lifshitz critical point, and the bending of the Lifshitz line near the two-phase region. The criticality is quantitatively characterized by the susceptibility critical exponent γ which shows a crossover from Ising to isotropic Lifshitz critical behavior (Fig. 26). There are a few distinctions between both systems studied. In the (PEE;PDMS) mixture the Lifshitz critical exponent of $\gamma = 1.62$ is observed over a plateau of about 2% diblock concentration. In the (dPB;PS) blend, γ depends on the defined reduced temperature field. Use of the conventional reduced temperature leads to a $2\gamma = 3.24$ at the DCP, while use of the modified reduced temperature field, namely $\hat{t} = (1 - T_{UC}/T)(1 - T_{LC}/T)$, which is defined by the UCST and LCST critical temperatures T_{UC} and T_{LC}, leads to $\gamma = 1.62$ at the DCP, which is the same as for the other blend (Fig. 26). The disordered phases above and below the two-phase region must be different as the driving force for phase decomposition is of enthalpic and entropic origins, respectively; the disordered phase at lower temperature is represented by a droplet microemulsion phase which must be the more ordered state in comparison with the other two phases.

A principle message from this article should be that thermal composition fluctuations play an essential and complex role in polymer blends. This was demonstrated by the crossover phenomena between mean field to 3D-Ising, renormalized 3D-Ising, isotropic Lifshitz, and renormalized isotropic Lifshitz critical behavior. The observation of such crossover phenomena may give hints to microstructural effects such as density fluctuations for the renormalized 3D-Ising case. The range of relevant thermal fluctuations in the Ising case is sensitively determined by the entropic term of the FH parameter. So, in order to accurately determine the FH parameter from scattering experiments the susceptibility must be analyzed in terms of a crossover function describing the transition from mean field to 3D-Ising behavior.

The FH parameter is a phenomenological parameter describing the phase behavior of polymer blends and is determined from the structure factor at $Q = 0$, thereby representing an average over the whole sample volume. Information about the microstructural origin of the FH parameter can be achieved from exploration of the blends under different conditions of molar volume, chain microstructure, additives, pressure and temperature fields. In this respect, the investigation of simple model systems is most promising as demonstrated by the SANS studies of the dPB/PB copolymer mixtures with different vinyl contents and molar volumes (Figs. 9 and 10). In this blend several microstructural contributions to the FH parameter could be separated, such as the chain-end effect, the polarization effect due to deuteration, and finally the enthalpic and entropic terms of the FH parameter could be fully analyzed in terms of the van der Waals interaction energy and a non-combinatorial entropic expression from the lattice cluster theory calculations, respectively (Fig. 10). Binary blends of chemically different chains quite naturally lead to a more complex situation which cannot be analyzed so easily. But some microstructural aspects can be visualized for the blend-solvent systems in large pressure fields showing an interrelation between cavities and solvent molecules.

The presented experimental studies and corresponding theoretical background on fluctuation phenomena in polymer blends have shown a broad field of exploration. Many of the observed phenomena are not yet understood and could only be explained qualitatively. Hence, this article will hopefully help to stimulate further experimental and theoretical interest in this field.

Acknowledgements I would like to sincerely thank Dr. Henrich Frielinghaus and Dr. Vitaliy Pipich for their cooperative work over the recent years and for critically reading the manuscript. Finally, I would like to express my thanks to Prof. Karl F. Freed for carefully reading the manuscript and for his several comments.

References

1. Utracki LA (1989) Polymer Alloys and Blends. Hanser, München
2. Flory PJ (1971) Principles of Polymer Chemistry. Cornell University Press, Ithaca, New York
3. deGennes PG (1979) Scaling Concepts in Polymer Physics, chap x. Cornell University Press, Ithaca, New York
4. Binder K (1994) Adv Polym Sci 112:181
5. Schwahn D, Mortensen K (2000) Thermal Composition Fluctuations in Polymer Blends studied with Small Angle Neutron Scattering. In: Brown W, Mortensen K (eds) Scattering in Polymeric and Colloidal Systems. Gordon & Breach Science Publishers, Amsterdam
6. Higgins JS, Benoit H (1994) Polymers and Neutron Scattering. Clarendon Press, Oxford
7. Roe RJ (2000) Methods of X-Ray and Neutron Scattering in Polymer Science. Oxford University Press, New York, Oxford
8. Koningsveld R, Chermin HAG, Gordon M (1970) Proc R Soc London Ser A 319:331
9. For example: Uzunov DI (1993) Theory of Critical Phenomena. World Scientific, Singapore
10. v. Sengers J (1994) In: Kiran E, Sengers J, Levelt MH (eds) Supercritical fluids: Fundamentals for application. Kluwer Academic Publishers, Dordrecht
11. Lifshitz M , Dudowicz J, Freed KF (1994) J Chem Phys 100:3957
12. Dudowicz J, Lifshitz M, Freed KF, Douglas JF (1993) J Chem Phys 99:4804
13. Fredrickson GH, Ganesan V, Drolet F (2002) Macromolecules 35:16
14. Wang ZG (2002) J Chem Phys 117:481
15. Bates FS, Fredrickson GH (1994) Macromolecules 27:1065; Fredrickson GH, Liu AJ, Bates FS (1994) Macromolecules 27:2503
16. Sariban A, Binder K (1987) J Chem Phys 86:5859
17. Deutsch HP, Binder K (1993) J Phys II (France) 3:1049
18. Schwahn D, Mortensen K, Yee-Madeira H (1987) Phys Rev Lett 58:1544
19. Schwahn D, Mortensen K, Springer T, Yee-Madeira H, Thomas R (1987) J Chem Phys 87:6078
20. Bates FS, Rosedale JH, Stepanek P, Lodge TP, Wiltzius P, Fredrickson GH, Hjelm RP (1990) Phys Rev Lett 65:1893
21. Hair DW, Hobbie EK, Douglas J, Han CC (1992) Phys Rev Lett 68:2476
22. Hair DW, Hobbie EK, Nakatani AI, Han CC (1992) J Chem Phys 96:9133
23. Chou B, Ying Q, Linliu K, Xie P, Gao T, Li Y, Nose T, Okada M (1992) Macromolecules 25:7382
24. Janßen S, Schwahn D, Springer T (1992) Phys Rev Lett 68:3180
25. Meier G, Momper B, Fischer EW (1992) J Chem Phys 97:5884
26. Theobald W, Meier G (1995) Phys Rev E 51:5776
27. Belyakov MY, Kiselev SB (1992) Physica A 190:75
28. Anisimov MA, Kiselev SB, Sengers JV, Tang S (1992) Physica A 188:487

29. Anisimov MA, Povodyrev AA, Kulikov VD, Sengers JV (1995) Phys Rev Lett 75:3146
30. Anisimov MA (2000) J Phys: Condens Matter 12:A451
31. Schwahn D, Meier G, Mortensen K, Janßen S (1994) J Phys II 4:837; Meier G, Schwahn D, Mortensen K, Janßen S (1993) Europhys Lett 22:577
32. Schwahn D, Frielinghaus H, Willner L (2002) J Chem Phys 116:2229
33. Patterson D, Robard A (1978) Macromolecules 11:690
34. Janßen S, Schwahn D, Mortensen K, Springer T (1993) Macromolecules 26:5587
35. Hammouda B, Bauer BJ (1995) Macromolecules 29:4505
36. Schwahn D, Schmackers T, Mortensen K (1995) Phys Rev E 52:R1288
37. Frielinghaus H, Schwahn D, Willner L (2001) Macromolecules 34:1751
38. Schwahn D, Frielinghaus H, Mortensen K, Almdal K (2001) Macromolecules 34:1694
39. Frielinghaus H, Schwahn D, Dudowicz J, Freed KF, Foreman KW (2001) J Chem Phys 114:5016
40. Frielinghaus H, Schwahn D, Willner L, Freed KF (2002) J Chem Phys 116:2241
41. Lefebvre AA, Lee JH, Balsara NP, Hammouda B, Krishnamoorti, Kumar S (1999) Macromolecules 32:5460
42. Beiner M, Fytas G, Meier G, Kumar SK (1998) Phys Rev Lett 81:594; (2002) J Chem Phys 116:1185
43. Schwahn D, Frielinghaus H, Mortensen K, Almdal K (2001) Macromolecules 34:1694
44. Miyashita N, Nose T (1995) Macromolecules 28:4433
45. Nose T, Miyashita N (2000) Critical Behavior in Polymer Blend Solutions. In: Brown W, Mortensen K (eds) Scattering in Polymeric and Colloidal Systems. Gordon & Breach Publishers, Amsterdam
46. Bates FS, Maurer W, Lodge TP, Schulz MF, Matsen MW, Almdal K, Mortensen K (1995) Phys Rev Lett 75:4429
47. Bates FS, Maurer WW, Lipic PM, Hillmyer MA, Almdal K, Mortensen K, Fredrickson GH, Lodge TP (1997) Phys Rev Lett 79:849
48. Schwahn D, Mortensen K, Frielinghaus H, Almdal K (1999) Phys Rev Lett 82:5056
49. Schwahn D, Mortensen K, Frielinghaus H, Almdal K, Kielhorn L (2000) J Chem Phys 112:5454
50. Hillmyer MA, Maurer WW, Lodge TP, Bates FS, Almdal K (1999) J Phys Chem 103:4814
51. Morkved TL, Chapman BR, Bates FS, Lodge TP, Stepanek P, Almdal K (1999) Faraday Discuss 112:335
52. Pipich V, Schwahn D, Willner L (2004) Physica B 350:e897
53. Brazovskii SA (1975) JETP 41:8552
54. Holyst R, Schick M (1992) J Chem Phys 96:7728
55. Broseta D, Fredrickson GH (1990) J Chem Phys 93:2927
56. Kielhorn L, Muthukumar M (1997) J Chem Phys 107:5588
57. Kudlay A, Stepanow S (2002) Macromol Theory Simul 11:16
58. Alefeld B, Schwahn D, Springer T (1989) Nucl Inst in Physical Research A 274:210
59. Schwahn D, Meier G, Springer T (1991) J Appl Cryst 24:568
60. Wignall GD, Bates FS (1987) J Appl Cryst 20:28
61. Schwahn D, Hahn K, Streib J, Springer T (1987) J Chem Phys 93:8383
62. Tabulated values in (1992) Neutron News 2:29
63. Flory PJ (1970) Disc Faraday Soc 49:7
64. Schwahn D, Willner L (2002) Macromolecules 35:239
65. Schwahn D, Mortensen K, Janssen S (1994) Phys Rev Lett 73:1452
66. Le Guillon JC, Zinn-Justin J (1992) Phys Rev Lett 39:95; Zinn-Justin J (1981) J Physique 42:783; LeGuillon JC, Zinn-Justin J (1987) J Physique 48:19

67. Melnichenko YB, Anisimov MA, Povodyrev AA, Wignall GD, Sengers JV, Van Hook WA (1997) Phys Rev Lett 79:5266
68. Melnichenko YB, Wignall GD, Schwahn D (2002) Phys Rev E 65:061802
69. ten Brinke G, Karasz FE, MacKnight WJ (1983) Macromolecules 16:1827
70. Paul DR, Barlow JW (1984) Polymer 25:487
71. Dudowicz J, Freed KF (2000) Macromolecules 33:3467
72. Dudowicz J, Freed KF (2000) Macromolecules 33:9777
73. Bates FS, Wignall GD, Koehler WC (1985) Phys Rev Lett 55:2425
74. Graessley WW, Krisnamoorti K, Balsara NP, Fetters LJ, Lohse DJ, Schulz DN, Sissano JA (1993) Macromolecules 26:1137
75. Jinnai H, Hasegawa H, Hashimoto T, Han CC (1992) Macromolecules 25:6078
76. Tang H, Freed KF (1991) Macromolecules 24:958
77. Coleman MM, Serman CJ, Bhagwagar DE, Painter PC (1990) Polymer 31:1187
78. For example: Ma SK (1982) Statistical Mechanics. World Scientific, Singapore, p 264
79. Fredrickson GH, Leibler L (1989) Macromolecules 22:1238
80. Broseta D, Leibler L, Joanny JF (1987) Macromolecules 20:1935
81. Schwahn D, Takeno H, Willner L, Hasegawa H, Jinnai H, Hashimoto T, Imai M (1994) Phys Rev Lett 73:3427
82. Anisimov A, Voronel AV, Gorodetskii EE (1971) Sov Phys JETP 33:605
83. Fisher ME (1968) Phys Rev 176:257
84. Pipich V, Schwahn D, Willner (2005) Phys Rev Lett 94:117801
85. Müller M, Gompper G (2002) Phys Rev E 66:041805
86. Nicoll JF, Tuthill GF, Chang TS, Stanley HE (1977) Physica 86–88B:618
87. Kielhorn L, Muthukumar M (1999) J Chem Phys 110:4079
88. Narayanan T, Kumar A, Gopal ESR (1990) Phys Lett A144:371; Prafulla BV, Narayanan T, Kumar A (1992) Phys Rev A 45:1266

Adv Polym Sci (2005) 183: 63–126
DOI 10.1007/b135883

Published online: 26 July 2005

Influence of Monomer Molecular Structure on the Miscibility of Polymer Blends

Karl F. Freed (✉) · Jacek Dudowicz

James Franck Institute and Department of Chemistry, The University of Chicago, Chicago, IL 60637, USA
k-freed@uchicago.edu

Abstract Polymer blends are formulated by mixing polymers with different chemical structures to create new material with beneficial properties of the individual components. While Flory–Huggins (FH) theory explains some basic trends in blend miscibility, the theory completely neglects the dissimilarity in monomer structures that is central to the fabrication of real blends. We systematically investigate the influence of monomer structure on blend miscibility using the lattice cluster theory (LCT) generalization of the FH model in the limit of incompressible, high molecular weight blends where analytical calculations are tractable. The well-known miscibility pattern predicted by FH theory is recovered only for a limited range of monomer size and shape asymmetries, but additional contributions to the LCT entropy and internal energy of mixing for polymers with dissimilarly shaped monomers leads to three additional blend miscibility classes whose behavior is quite different from the prediction of classical FH theory. Several illustrative applications of the LCT provide new molecular-scale interpretations for many nontrivial phenomena occurring in polymer systems. The applications also illustrate the predictive ability of the theory and its usefulness in analyzing thermodynamic data for a wide variety of polymer mixtures, ranging from binary homopolymer blends to various types of copolymer systems.

Keywords Polymer blends · Miscibility · Statistical copolymers · Nonrandom mixing effects · Monomer structural asymmetry

1 Introduction

Despite the fact that the production of polymers (including the preparation of monomers) accounts for roughly half of the chemical industry in the US, polymer science represents a minuscule portion of the education of chemistry students here. Nevertheless, the fields of polymer physics and polymer chemistry afford great scientific and technological interest. Polymer systems sustain a wide variety of phases, and materials fabricated from the different phases may exhibit enormously varying properties. A number of challenging conceptual questions concern phase separation in polymer blends, phase transitions in liquid crystals and gels, and the glass transition in polymer systems. For instance, many commercially important materials are *alloys* of polymers having different chemical and physical characteristics, and the stability and state of dispersion of polymer blends are often crucial in applications. Thus, it is both a scientific challenge and of technological importance to elucidate the molecular features promoting the mixing of polymers and governing the compositions of the coexisting phases when polymers do not mix. This challenge is heightened by the recent development [1, 2] of metallocene catalysts for polyolefin production, which has revolutionized the tailoring of polyolefins into thermoplastic elastomers and other improved materials. This progress in polyolefin synthesis creates a demand for a structure-property theory to guide the syntheses. The utility of theory for designing novel advanced materials depends, to a large extent, on its ability to describe the essential molecular factors determining the system's thermodynamics and consequently its phase behavior.

The overwhelming majority of thermodynamic treatments for multicomponent polymer systems are based on Flory–Huggins (FH) [3–7] theory and various embellishments thereof [8–16]. While these theories are extremely useful in correlating data from measurements of different properties and in explaining many facets of the miscibility of polymer blends, the theories suffer from several significant deficiencies that limit their predictive abilities and, in particular, their utility for tackling many technologically important issues [17]. The limitations of prior theories are discussed in Sect. 1.1 to emphasize the scientific challenges and opportunities open to theories that are devoid of these deficiencies and to accentuate the motivation for the development of the lattice cluster theory (LCT) [17–22], which is the subject of the present review. Several illustrative applications of the LCT provide new molecular-scale interpretations for many enigmatic phenomena occurring in polymer systems. The applications further demonstrate the predictive potential of the theory and its usefulness in analyzing thermodynamic data for a wide variety of systems, ranging from binary homopolymer blends to various types of copolymer mixtures.

1.1 Critical Review of Flory–Huggins Theory: Motivation for Developing Improved Analytical Theory

FH theory [3–7] was originally derived from a simple lattice model that imposes a discrete spatial representation of the conformations accessible to individual polymer chains and, *per force*, to collections of many intertwined polymers in the liquid phase. The polymers are taken to have random configurations, which arise from placing individual monomers at single lattice sites, subject to strict constraints of single occupancy for any lattice site. The same single occupancy constraint applies to solvent molecules when they are also present. The exact counting of configurations for a dense system of flexible polymers is beyond feasibility, so FH theory [3–7] employs a very simple mean-field approximation that effectively ignores the details of the polymer chain connectivity and, therefore, that cannot distinguish between linear, star, branch, comb, etc. polymer architectures.

The basic physical content of FH theory [3–7] is embodied in its treatment for the entropy and enthalpy of mixing. In this regard, polymers behave quite differently from small molecules. The large mixing entropy in mixtures of small molecules is generally responsible for their miscibility in the liquid phase because the enthalpy of mixing is frequently unfavorable. The entropy of mixing for small molecule fluids is proportional to the familiar $\sum_i n_i \ln(n_i)$ term (called the combinatorial entropy), where n_i denotes the number of molecules of species i in the system. As extended to large irregular shaped objects such as polymers, FH theory replaces the $\sum_i n_i \ln(n_i)$ term by the expression $\sum_i n_i \ln(\varphi_i)$, where φ_i is the volume fraction for polymer species i. The large number of monomers present in individual polymer chains implies that a given mass of a polymer blend contains a rather small number of molecules compared, for instance, to the same mass of a monomer mixture. Consequently, the entropy of mixing is small, and the entropic force driving polymer miscibility is weak.

FH theory [3–7] employs a random mixing (regular solution) approximation for the energy of mixing. More specifically, if ε_{ij} is the van der Waals attraction energy between monomers of species i and j that are placed at neighboring lattice sites, and if the exchange energy is defined as $\varepsilon = \varepsilon_{ii} + \varepsilon_{jj} - 2\varepsilon_{ij}$, the dimensionless energy of mixing for an incompressible binary mixture is equal to the product of volume fractions $\varphi_i \varphi_j$ and the effective Flory interaction parameter,

$$\chi = z\varepsilon/2kT\,, \tag{1}$$

where kT is the thermal energy and z is the number of lattice sites that are nearest neighbors to a given lattice site. For interactions dominated by dispersion forces, the χ parameter is expected to be positive, so the occurrence of a positive χ and a small entropy of mixing explains why polymers of different species generally do not mix in the liquid phase. However, some polymer species do mix well. Thus, many technologically important materials are fabricated from mixtures of different polymers, and the resultant composite materials often exhibit a synergistic combination of beneficial characteristics of the individual blended components. For instance, blending a soft rubbery polymer and a brittle tough polymer may yield a strong, elastic substance. Consequently, a greater understanding of the factors governing blend miscibility is of technological interest, and the present review illustrates how these insights into polymer blend miscibility can emerge from theories that describe how the χ parameter depends on the detailed monomer molecular structure and the intermolecular interactions.

The FH expression for χ in Eq. 1 provides little further insight into the physical features that control polymer miscibility in view of the fact that typical values of χ for miscible polymers are only of the order of a few Kelvin, while typical values for the individual ε_{ij} are hundreds of Kelvin, leading to a low likelihood for computing the χ parameter from first principles. Consequently, the χ parameter is traditionally treated in a phenomenological fashion. Nevertheless, the theory has been immensely useful in correlating data from different types of measurements, including those associated with the miscibility of polymer blends. Hence, studying the behavior of χ for a set of blends formed from a homologous series of polymers should assist in identifying the factors governing the thermodynamic behavior of polymer systems. Indeed, a large body of empirical data for a series of polyolefin blends exhibits glaring discrepancies from the predictions of FH theory.

Inspection of Eq. 1 indicates that χ is *predicted* by FH theory [3–7] as inversely proportional to temperature and as independent of polymer composition (the φ_i), molecular weights, and pressure. However, χ is often found empirically to depend on the blend composition, sometimes to vary with molecular weights, and to be a function of pressure. The pressure dependence of χ has been observed in several experiments [23, 24] subsequent to our predictions [21] that χ is not insensitive to pressure, and the experiments show that the theory correctly estimates [25, 26] the magnitude of this technologically important dependence. Moreover, as *reviewed* by Flory [12] over 30 years ago, the temperature dependence of χ generally exhibits the form $\chi = A + B/T$, with the presence of a temperature-independent portion A, called the "entropic" part of χ, standing in sharp contrast to the predictions of FH theory [3–7], which rigorously omits the A term. These strong contradictions between theory and experiment *logically* imply (as noted by Flory [12]) that FH theory is *wrong* concerning the predicted behavior of χ (i.e., *the theory is grossly inadequate to explain general observations for* χ). Thus, treating χ as a phenomenological parameter and allowing this empirical χ to display the observed variations (with temperature, composition, pressure, etc.) *technically* represents pure empiricism (as Flory recognized) and not consistent use of FH theory, *per se*. While this empiricism is perfectly reasonable for analyzing experimental data, the apparent successes of this empiricism cannot overshadow the lack of theoretical underpinnings that are necessary for more ambitious predictive purposes than just correlating data for a specific system. All mention of FH theory below refers to the original definition in Eq. 1 and not to the phenomenological treatment of χ that effectively is a tautology.

The above-mentioned limitations to FH theory [3–7] are merely the tip of the iceberg when questions are raised concerning the molecular design of new materials. Within a lattice model, FH theory specifies, for instance, that monomers of different polymer species and solvent molecules (if present) all occupy single-lattice sites, i.e., have the same shapes and volumes. Realists may argue that actual differences in monomer volumes are maintained by the theory if different numbers of monomers of different species are taken to reside at a given lattice site. However, this assumption still grossly ignores significant details of monomer structure, such as disparate monomer molecular sizes and *shapes* and, consequently, the influences of these structural differences on the thermodynamics of polymer systems. Thus, a fundamental question posed in formulating an analytical molecular-based theory for the statistical thermodynamics of liquid-phase polymers is associated with recognizing the degree to which the distinct chemical structures of the individual monomers are relevant. This question represents a substantial departure from the traditional strategy used in theoretical polymer physics. The latter theoretical treatments focus on *universal* polymer properties that arise intrinsically from polymer chain connectivity and that are consequently independent of the detailed chemical nature of the system, apart from a few empirical parameters, such as the χ parameter discussed above. Our molecular-based theory, instead, relates these empirical parameters, and particularly the χ parameter, to specific chemical characteristics of individual monomers and solvent molecules and to the thermodynamic state of the system.

1.2 Strategies for Developing Improved Analytical Theories

A considerable body of recent theoretical analyses for polyolefin blends [27–32] attempts to relate the χ parameter to differences $(\delta_1 - \delta_2)^2$ in solubility parameters, where $(\delta_i)^2$ is proportional to the cohesive energy of the pure component i (i.e., to the difference between the internal energies of the pure liquid and vapor). This cohesive energy is often approximated by the more accessible internal pressure $(\partial U/\partial V)_T$ and, therefore, does not contain contributions from the athermal packing entropy that is responsible for the appearance of an important component of the entropic χ parameter for many polymer systems. Hence, when the empirical χ parameter contains a significant temperature-independent portion χ_s, solubility parameter theory can at most only include χ_s as a theoretically unjustified, purely empirical quantity, just as in traditional applications of FH theory. Moreover, many polymer solubility parameters are determined from thermodynamic data for polymer solutions where the entropic component of χ_s may be dominant [12]. Therefore, this type of empiricism contradicts the *fundamental assumption* of solubility parameter theory that χ is of energetic origin. Thus, a major conceptual extension of solubility parameter theory would involve incorporating the entropic χ in a theoretically correct fashion and relating both the entropic and energetic contributions of the χ parameter to the monomer chemical structures and their interactions.

A number of analytical off-lattice approaches are based on the extension of various simple one-component analytical models (such as Flory–Huggins theory, Flory's equation of state model, or Guggenheim theory) to multicomponent systems by the use of some ad hoc *combining rules* [5–8, 15, 16]. These approaches likewise neglect the influence of chain connectivity and monomer structure on local nonrandom mixing effects whose many ramifications we wish to understand. While improving the predictive abilities of analytical theories, it is also desirable simultaneously to develop the numerically intensive off-lattice integral equation methods that are capable of describing the properties of polymer systems that contain structured monomers, of course, at the expense of heavy numerical computations. (See, for example, recent numerical studies of polyolefin melts using PRISM theory [33–35].)

Another strategy for developing molecular-based theories consists in generating systematic corrections to FH theory that include details of monomer structure and describe the influence of short range correlations [17–22]. While FH theory may be derived without invoking a lattice model, it is traditionally regarded as a lattice model approach because of its original formulation. Since the traditional FH lattice model treats the monomers as structureless units, a first step in this strategy of systematically improving FH theory involves replacing the FH model with a modified (extended) lattice model in which monomers are represented as structured entities. Thus, the monomers are permitted to occupy several neighboring lattice sites to mimic the sizes, shapes, and structures of the actual molecules. Although chemists and chemical engineers often view lattice models as providing an oversimplified representation of reality, physicists find these models to be extremely attractive since their use introduces enormous mathematical tractability while still preserving the essential physics of the system. Because the simple mean-field approximations inherent to FH theory [3–7] do not distinguish between different monomer structures, we have developed [18–22] a vastly superior, approximate analytical solution to the lattice model of dense polymer systems. The resulting approach is called the lattice cluster theory to emphasize similarities with the Mayer cluster theory of nonideal gases [36].

2
Theoretical Basis for the Lattice Cluster Theory

The lattice cluster theory (LCT) [18–22] is based on an extended lattice model where monomers are endowed with molecular structures, as illustrated in Fig. 1a and b, which depict united atom group structures for the monomers of several polyolefins and portions of a few polyolefin chains, respectively. Individual CH_n ($n = 0–3$) groups are taken as occupying single lattice sites, and the bonds between united atom groups correspond to the C – C bonds in the actual molecules. The terminal segments of the side groups in Fig. 1a and b are methyl (CH_3) groups, while the backbone united atom groups represent CH_n units, with $n = 2$, 1, or 0. These united atom models have long been applied in off-lattice theories and in computer simulations of linear and

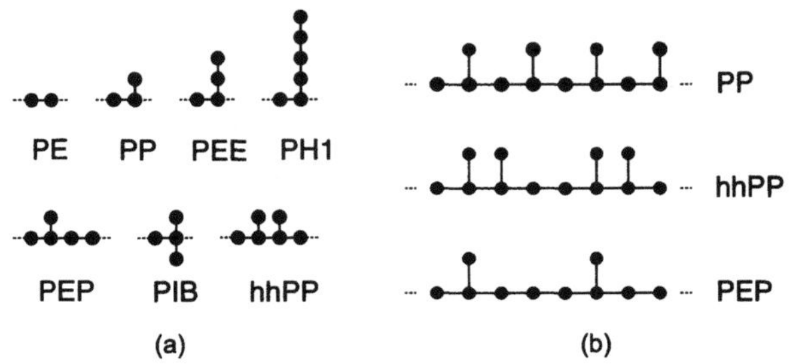

Fig. 1 (**a**) United atom group models for monomers of poly(ethylene) (PE), poly(propylene) (PP), poly(ethyl ethylene) (PEE), poly(hexene-1) (PH1), poly(ethylene propylene) (PEP), Poly(isobutylene) (PIB), and head-to-head poly(propylene) (hhPP). *Circles* designate CH_n groups, *solid lines* represent C – C bonds inside the monomer, and *dotted lines* indicate the C – C bonds linking the monomer to its neighbors along the chain. (**b**) United atom models of PP, hhPP, and PEP homopolymer chains constructed by linking monomers of (**a**)

branched alkanes [37–40], perhaps the simplest of all homologous molecular systems.

FH theory [3–7] neglects all correlations that are present in the system due to bonding and excluded volume constraints and to the van der Waals interactions. The LCT, on the other hand, describes the thermodynamic consequences of the existence of short-range correlations that arise from packing constraints and the different monomer–monomer interactions. Because only short-range correlations are included, the LCT technically is also a mean-field theory. It is clear that the inclusion of these correlations in a theory of polymer systems requires developing fundamentally new theoretical concepts and counting methods. The goal of the LCT is to determine the influence of monomer structure on the thermodynamic properties of specific polymer systems. The basic physical underpinnings of the lattice cluster theory are sketched in the remainder of this section, and the interested reader may consult the original literature for the lengthy, complicated details of the theory [18–22]. Those more interested in applications may skip ahead to Sect. 3 and the following sections, which describe several predictions of a simplified version of the LCT. In these sections, we analyze basic phenomena whose explanation lies outside the scope of simple FH theory.

2.1 Brief Description of Basic Theoretical Principles

The partition function for the extended lattice model of polymer systems is evaluated within the LCT by introducing approximations for the bonding constraints and for the excluded volume and van der Waals interactions. For simplicity, we assume for now that all chains are fully flexible and focus first on the description of the partition function Z_{pack} for a single athermal limit polymer chain on a hypercubic lattice (a cubic lattice in d dimensions, so $z = 2d$). Generally, the chain connectivity constraint (i.e., the condition that successive monomers are linked to each other) is represented in Z_{pack} by a product of the Boltzmann factors

$$b_i = \exp[-U_{\text{bond}}(\boldsymbol{r}_{i-1} - \boldsymbol{r}_i)/kT], \tag{2}$$

where $U_{\text{bond}}(\boldsymbol{r}_{i-1} - \boldsymbol{r}_i)$ is the potential energy for the chemical bond between the united atom units, labeled $i - 1$ and i, in the polymer and kT is the absolute temperature in energy units. Each of these Boltzmann factors b_i simplifies enormously in the lattice model to the condition that the vector $\boldsymbol{r}_{i-1} - \boldsymbol{r}_i$ connecting units $i - 1$ and i is one of the z possible lattice vectors $\boldsymbol{a}_\beta(\beta = 1, \ldots, z)$ between nearest-neighbor sites on the lattice, where z is the number of nearest neighbors to any lattice site. The above simplification may be applied to *each* bonding interaction as

$$\exp[-U_{\text{bond}}(\boldsymbol{r}_{i-1} - \boldsymbol{r}_i)/kT] \to \sum_{\beta_{i-1}=1}^{z} \delta(\boldsymbol{r}_{i-1} - \boldsymbol{r}_i - \boldsymbol{a}_{\beta_{i-1}}), \tag{3}$$

where δ is the Kronecker delta function,

$$\begin{aligned} \delta(i,j) &= 1 \quad \text{if} \quad i = j, \\ &= 0 \quad \text{otherwise}, \end{aligned}$$

and the sum over β appears because the bond may lie along any of z possible directions.

The approximate treatment of the bonding constraints in Eq. 3 may be motivated by recourse to the *Flory theorem* [3, 4], which states that in polymer melts it is impossible to discern whether a pair of nonbonded nearest-neighbor united atom groups belongs to different polymer chains or to distant portions of the same polymer molecule. Thus, in the lattice model description of polymer systems, the excluded volume prohibition of multiple occupancy of a site is more important than the consequences of long-range chain connectivity. Based on the Flory theorem, we introduce the zeroth-order mean-field average

$$A = \langle \delta(\boldsymbol{r}_{i-1} - \boldsymbol{r}_i - \boldsymbol{a}_{\beta_{i-1}}) \rangle$$

to reexpress the δ on the right-hand side of Eq. 3 exactly as

$$\delta(\boldsymbol{r}_{i-1} - \boldsymbol{r}_i - \boldsymbol{a}_{\beta_{i-1}}) \equiv zA(1 + X_{i-1,i}), \tag{4}$$

with the bond correlation factor $X_{i-1,i}$ given by

$$X_{i-1,i} = (A)^{-1}[\delta(\boldsymbol{r}_{i-1} - \boldsymbol{r}_i - \boldsymbol{a}_{\beta_{i-1}}) - A]. \tag{5}$$

The overall chain connectivity Boltzmann factor for a single polymer chain with specified positions $\{\boldsymbol{r}_i\}$ of the united atom groups is the product (over all

N bonds in the chain) of individual terms in Eq. 3 and can be written using Eqs. 4 and 5 as

$$\prod_{i=1}^{N} \exp[-U_{\text{bond}}(\boldsymbol{r}_{i-1}-\boldsymbol{r}_i)/kT] = \prod_{i=1}^{N} \sum_{\beta_{i-1}}^{z} A(1+X_{i-1,i}) \,. \tag{6}$$

The athermal limit partition function Z_{pack} for packing a system of n monodisperse purely flexible linear chains is computed by constructing the product of factors from the right-hand side of Eq. 6 for each polymer chain in the system and then by summing the resultant expression over all possible positions $\{\boldsymbol{r}_i\}$ of the united atom units, subject to the strict excluded volume constraints that no two segments may occupy the same lattice site. Thus, the athermal packing partition function Z_{pack} takes the form

$$Z_{\text{pack}} = \frac{1}{2^n n!} \sum_{\boldsymbol{r}_1^0 \neq \boldsymbol{r}_2^1 \neq \ldots \neq \boldsymbol{r}_n^N} \prod_{c=1}^{n} \prod_{i=1}^{N} \sum_{\beta_{i-1}^c=1}^{z} [A(1+X_{i-1,i}^c)] \,, \tag{7}$$

where the index $c = 1, \ldots, n$ labels the chains, $i = 1, \ldots, N$ labels the bonds between monomers i and $i-1$ within a chain, and the factors of 2^{-n} and $1/n!$ account, respectively, for the indistinguishability of the two chain ends and of the chains themselves.

The product in Eq. 7 generates a cluster expansion for the partition function Z_{pack} that bears some similarity to the Mayer cluster expansion [36], which is used in the theoretical description of nonideal gases. The average quantity A is defined such that the leading zeroth-order contribution to Z_{pack} recovers the Flory combinatorial entropy [the $\sum_s n_s \ln(\varphi_s)$ term, where φ_s is the volume fraction of species s], while the $X_{i-1,i}$ terms represent nonrandom mixing corrections, including a temperature-independent contribution to the χ parameter that arises from the correlations introduced by the chain connectivity and excluded volume constraints. The free energy F_{pack} for a mixture of athermal, flexible chains is obtained from Eq. 7 by applying the standard relation

$$F_{\text{pack}} = -kT \ln Z_{\text{pack}}$$

and has the form of a series in powers of $1/z$.

Equations 3–7 have also been extended to describe nonathermal chains interacting with nearest-neighbor van der Waals energies $\{\varepsilon_{cd}\}$, where the subscripts c and d label the polymer species and all united atom groups of a given monomer are assumed, for simplicity, to interact with a common interaction energy. (Sect. 9 briefly sketches a straightforward generalization of the LCT to cases in which different interaction energies are ascribed to chemically distinct groups within the monomers.) The resultant free energy F emerges as a double expansion in $1/z$ and $\{\varepsilon_{cd}/kT\}$. The athermal and order ε_{cd} contributions have been evaluated through order $1/z^2$, while the order

$(\varepsilon_{cd})^2$ terms have been determined so far through order $1/z$. The truncation of the series in $\{\varepsilon_{cd}/kT\}$ implies that the LCT applies only in the range of higher *relative* temperatures. A significant benefit of the LCT lies in the algebraic nature of the expressions for thermodynamic properties that apply for all compositions, molecular weights, interaction energies ε_{ij}, monomer structures, and temperatures, with the only constraint that the $\{\varepsilon_{ij}/T\}$ not be *too large*.

The LCT excess free energy for a binary homopolymer blend in the incompressible (infinite pressure) limit depends on the interaction energies only through the single dimensionless exchange energy $\varepsilon/kT = (\varepsilon_{11} + \varepsilon_{22} - 2\varepsilon_{12})/kT$. Since ε/kT for polymer blends is typically of order 10^{-2} (or less) per united atom group, the LCT free energy series in ε/kT may safely be truncated at order $(\varepsilon)^2$ when invoking the simplifying incompressible limit discussed in Sect. 3.

2.2
Tests of the LCT

The accuracy of the LCT has been tested against Monte Carlo (MC) simulations of various lattice model polymer systems. Recent simulations by Buta et al. [41] generate MC data for several thermodynamic properties [chemical potential, entropy, internal energy, isothermal compressibility, and specific heat] of the linear chain polymer/solvent system for a wide range of temperatures and for polymer volume fractions up to 0.8. These simulations demonstrate that the LCT provides an excellent estimation of all thermodynamic properties in the mean-field regions, with the exception of the specific heat at lower temperatures. Errors are, of course, evident for dilute and semidilute systems where non-mean-field effects become important. These tests of the LCT have been performed without the use of any adjustable parameters and, therefore, enable distinguishing whether differences between theory and experiment, if any, are due to deficiencies in approximating the partition function for the lattice model system or to inherent deficiencies associated with the lattice model itself. Indeed, an earlier comparative analysis [42] quickly indicated the need for introducing monomer structures to explain and describe the presence of the entropic contribution to the χ parameter that is rigorously absent in FH theory. Subsequent MC simulations for melts [43] and blends [44] of polymers with structured monomers imply that the LCT still exhibits some deficiencies in determining the variations of thermodynamic properties of various polymer systems with specific monomer structures. The elimination of these deficiencies requires the inclusion in the free energy expression of higher-order contributions beyond those calculated so far (Sect. 2.1). The evaluation of these higher-order contributions is, however, nontrivial.

3 Simplifying High Molecular Weight, Incompressible Limit of the LCT (the SLCT)

The theoretical description of polymer thermodynamics at a fixed pressure requires treating the system as compressible. The system's compressibility is introduced into the extended lattice models by allowing for the presence of empty lattice sites, called voids. The voids *do not* represent a species of particles in the thermodynamic sense, nor do they have interactions; they are merely a simple bookkeeping device for keeping track of the excess free volume. When computations are performed for constant pressure systems, the excess free volume fraction is determined numerically from the equation of state, derived as usual by defining the pressure as the derivative of the free energy with respect to total volume. The LCT expressions for the free energy of polymer melts, blends, and solutions are lengthy because local correlations are treated in a systematic, perturbative fashion [20, 21]. Nevertheless, the fact that the algebraic, analytical expressions apply for all compositions, monomer structures, molecular weights, temperatures, and interaction energies ε_{ij} (provided the ratios ε_{ij}/kT are sufficiently small) is a significant benefit of the LCT. The analytical tractability of the LCT contrasts with the numerical complexity of other theories, such as PRISM theory [45], which involves the use of special numerical techniques to solve a set of integral equations separately for each system studied (i.e., for each temperature, composition, etc.). Moreover, in certain limits the analytical expressions of the LCT become enormously compact and numerically tractable. For example, the lengthy LCT expression for the free energy of binary polymer blends reduces in the incompressible, high molecular weight limit to a simple, physically transparent form that provides great insights on the molecular scale into the behavior of a wide variety of polymer phenomena and systems. This particular version of the LCT is called the simplified lattice cluster theory (SLCT) [46–48] and is described below after a discussion of how the monomer structures are specified within the theory.

3.1 Description of Monomer Structures in the LCT

The dependence on monomer molecular structures enters the LCT free energy through a series of geometrical indices [18, 20] $N_i^{(\alpha)}$ that enumerate the number of distinct sets of α sequential bonds in a single chain of species i. The only geometrical coefficients, appearing in the SLCT (and the LCT) expression for the free energy of mixing for a homopolymer blend, are the ratios [49]

$$r_i \equiv \frac{N_i^{(2)}}{M_i} \quad \text{and} \quad p_i \equiv \frac{N_i^{(3)}}{M_i},$$

where M_i is the number of united atom units in a polymer chain of species i. The quantity r_i may be represented more conveniently (using Euler relations [50]) in terms of the respective numbers $s_i^{(\text{tri})}$ and $s_i^{(\text{tetra})}$ of tri- and tetrafunctional united atom groups in a single monomer of species i as

$$r_i \equiv \frac{N_i^{(2)}}{M_i} = 1 + \frac{s_i^{(\text{tri})}}{s_i} + 3\frac{s_i^{(\text{tetra})}}{s_i}, \tag{8}$$

where s_i designates the number of united atom units in a monomer of species i and where the large M_i limit has been invoked [51, 52].

The technical details associated with the application of Eq. 8 are explained below using the first three polyolefin structures depicted in Fig. 1b as examples. None of these three polyolefins contains tetrafunctional groups, so $s_i^{(\text{tetra})}/s_i = 0$ for all of them. Because only one of the three united atom groups of the PP monomer ($s_{\text{PP}} = 3$) is trifunctional, $s_{\text{PP}}^{(\text{tri})}/s_{\text{PP}} = 1/3$ and $r_{\text{PP}} = 1 + 1/3 = 4/3$. The hhPP monomer contains six united atom groups, two of which are trifunctional, and, therefore, $r_{\text{hhPP}} = 1 + 2/6$, which equals r_{PP}. Similar counting for a PEP monomer yields $r_{\text{PEP}} = 6/5$.

While the evaluation of p_i can be simplified using Euler relations for certain classes of monomer structures, these Euler relations [50] do not apply, for example, to the three polyolefin chains depicted in Fig. 1b because they have short side chains. (Of the monomer structures in Fig. 1a, the Euler relations for p_i are valid only for PEE and PH1.) Therefore, the geometrical index $p_i \equiv N_i^{(3)}/M_i = n_i^{(3)}/s_i$ is evaluated by directly enumerating all sets of three sequential bonds ($n_i^{(3)}$) that traverse a monomer of species i. [46] summarizes the details of these calculations and tabulates values of p_i for several monomer structures, so we pass now to a consideration of how thermodynamic properties depend on r_i and p_i in the high molecular weight, incompressible system limit of the LCT.

3.2 Basic Thermodynamic Properties

The specific free energy of mixing for a binary polymer blend emerges from the SLCT as [46]

$$\frac{\Delta F_{\text{mix}}}{N_l kT} = \frac{\varphi}{M_1}\ln\varphi + \frac{1-\varphi}{M_2}\ln(1-\varphi) \tag{9a}$$
$$+ \varphi(1-\varphi)\left[\frac{(r_1-r_2)^2}{z^2} + \frac{\varepsilon}{kT}\left(\frac{z-2}{2} - \frac{1}{z}\Big[p_1(1-\varphi) + p_2\varphi\Big]\right)\right],$$

where N_l is the total number of united atom groups in the system, the site occupancy index M_i denotes the number of united atom groups in a single chain of species i, the blend composition variable $\varphi \equiv \varphi_1 = 1 - \varphi_2$ is chosen as the volume fraction for species 1, and $\varepsilon = \varepsilon_{11} + \varepsilon_{22} - 2\varepsilon_{12}$ is the blend exchange energy. The corresponding small angle neutron scattering (SANS) interaction parameter χ'_{SANS} is defined in terms of the free energy of mixing ΔF_{mix} of Eq. 9a by

$$2\chi'_{\text{SANS}} = \left[\frac{1}{M_1\varphi} + \frac{1}{M_2(1-\varphi)} - \frac{\partial^2[\Delta F_{\text{mix}}/N_l kT]}{\partial\varphi^2}\bigg|_{T,N_l} \right] \tag{9b}$$

and, therefore, is expressed per united atom group. The prime in χ'_{SANS} is used to distinguish between the theoretical LCT united atom group-united atom group parameter χ'_{SANS} and the experimental monomer-monomer interaction parameter χ_{SANS}. These two parameters are related to each other by

$$\chi_{\text{SANS}} = C\chi'_{\text{SANS}}\,, \tag{9c}$$

where the conversion factor C depends on the definition of the normalizing volume v_0 used to calculate χ_{SANS} from the experimental zero-angle scattering intensities. If v_0 is chosen as the geometric mean of the monomer volumes of the two blend components [i.e., if $v_0 = (v_1 v_2)^{1/2}$], C is given by

$$C = (s_1 s_2)^{1/2}\,, \tag{9d}$$

while an alternative choice [53] of $v_0 = [\varphi v_1 + (1-\varphi)v_2]$ implies the scaling

$$C = \frac{s_1 s_2}{s_1(1-\varphi) + s_2\varphi}\,. \tag{9e}$$

When $s_1 = s_2 = s$ (i.e., $v_1 = v_2$), Eqs. 9d and 9e coincide with each other.

Evaluating the derivative of the free energy in Eq. 9b converts χ_{SANS} into the rather compact expression

$$\frac{\chi_{\text{SANS}}}{C} = \left(\frac{r_1 - r_2}{z}\right)^2 + \left(\frac{\varepsilon}{kT}\right)\left(\frac{z-2}{2} + \frac{1}{z}\left\{p_1[1 - 3(1-\varphi)] + p_2[1 - 3\varphi]\right\}\right) + O\left(\left(\frac{\varepsilon}{kT}\right)^2\right)\,, \tag{10}$$

with the scaling factor given by Eq. 9d or 9e. The first term on the right-hand side of Eq. 10 is the athermal limit *entropic* portion of χ_{SANS}, which depends on the chemical structures of the two blend components. The appearance of this temperature-independent portion in Eq. 10 becomes more understandable by referring to Eq. 7 and to the observation that Z_{pack} of Eq. 7 may be expressed as the sum of the leading term (without any $X^c_{i-1,i}$ factors), which reproduces the FH combinatorial entropy $S^{\text{FH}}_{\text{comb}}$, and the remainder that accounts for corrections to $S^{\text{FH}}_{\text{comb}}$. These corrections are present

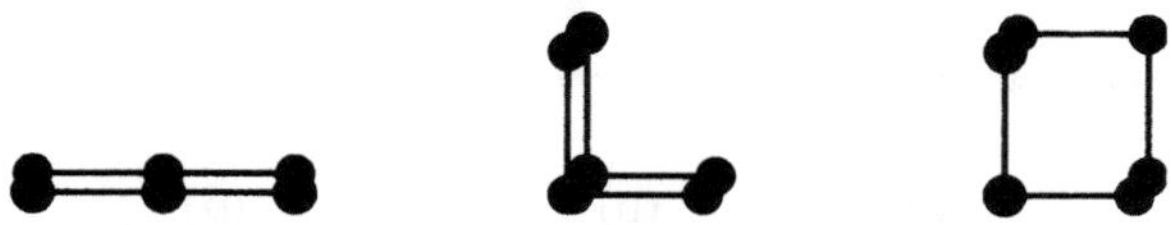

Fig. 2 Three possible overlapping configurations for two pairs of bonds on a two-dimensional square lattice

because the calculation of $S_{\text{comb}}^{\text{FH}}$ by Flory does not omit numerous classes of unphysical configurations in which excluded volume interactions are violated. Figure 2 displays examples of some disallowed configurations for pairs of two-bond sequences, configurations whose proper treatment contributes to the temperature-independent term in Eq. 10.

The two remaining terms in χ_{SANS} of Eq. 10 are of energetic origin and contain both monomer-structure-dependent and independent contributions. The leading energetic contribution of $z\varepsilon/2kT$ is merely the Flory–Huggins interaction term that grossly overestimates the number of nearest-neighbor heterocontacts [54, 55]. The replacement of the factor of z in the FH approximation $z\varepsilon/2kT$ by $z-2$ is consistent with the arguments of Guggenheim [56, 57] that each interior unit in a linear chain is linked by chemical bonds to two nearest-neighbor units. Consequently, these two neighboring sites are unavailable for occupancy by units belonging to the other species, thereby reducing the probability of heterocontact interactions. The second term of order ε/kT in Eq. 10 is a correction to the number of heterocontacts due to the packing constraints imposed by the monomer molecular structures. This last term is an explicit function of composition and the geometrical indices p_i. The composition-dependent portion of this term provides a correction to the random mixing approximation (in the strict probabilistic sense [58]). The $(\varepsilon/kT)^2$ contributions in Eq. 10 are generally negligible compared to the ε/kT in this equation. (As is well known, several definitions of theoretical χ_{SANS} parameters are used in the literature, and generally these definitions differ when χ_{SANS} depends on composition.)

The appearance of the counting indices p_i in the coefficient of ε in Eq. 10 represents a generalization of the Guggenheim *surface fraction* concept to structured monomers. This important observation is rendered more evident by considering the representative sequences of three consecutive bonds even on a linear chain that are depicted in Fig. 3. For any of the configurations in Fig. 2a–c, all neighboring sites to the four bonded united atom units can be occupied by a united atom unit of another polymer species. When these three bonds occur in a U-shaped conformation as in (**3d**), the two end units in the three-bond sequence are nearest neighbors to each other. Consequently, one less neighboring lattice site is available for possible heterocontact interactions. These considerations indicate that the previously vague concept of *surface fractions* [56, 57] is automatically quantified and extended to struc-

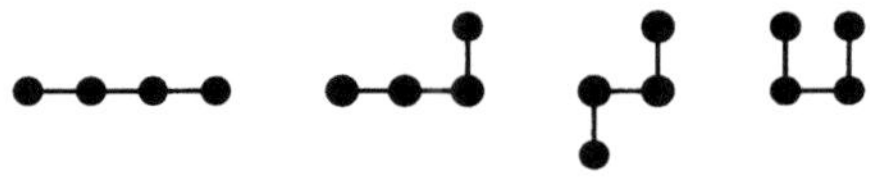

Fig. 3 Some planar configurations of three sequential bonds. All available neighboring sites to the bonded united atom groups in the configurations (**a**)–(**c**) may be occupied by united atom units of another species, while the end units in configuration (**d**) are nearest neighbors to each other, which effectively decreases the probability of heterocontact interactions with units of another species

tured monomer systems by the SLCT. Moreover, the properly determined surface fractions must be composition dependent, and Eq. 10, indeed, exhibits this dependence. The determination of the statistical distribution of contacts becomes more tedious for compressible systems because neighboring sites may be vacant or may be occupied by united atom units of either species. (The SLCT expression for the free energy of polymer solutions is similarly more complicated than Eq. 9a.)

4
Classes of Blend Miscibility Associated with Structural Asymmetries

Early LCT computations [22, 59] indicated that changes in monomer molecular structure alone could produce large alterations in the χ parameter and, hence, in the miscibility of polymer blends. Subsequent small angle neutron scattering (SANS) experiments for binary polyolefin blends by Graessley, Lhose, and their coworkers [27–32, 60] confirm these LCT predictions. In particular, these experiments reveal the existence of a wide range of miscibility patterns that contrast with the naive view of polyolefins as a homologous family of chemically similar compounds that should therefore be mutually miscible. The present and next three sections describe some SLCT calculations for binary homopolymer blends. As illustrated below, monomer structural asymmetry is found to affect profoundly not only the blend critical temperature and composition, but also the chain swelling and the scale and intensity of composition fluctuations [47, 48]. The SLCT predicts [47, 48] the existence of four qualitatively different categories of blend miscibility, only one of which can be explained by FH theory. Before characterizing these three additional categories, we note that retaining all contributions involving the excess free volume and all $1/M$ corrections in the LCT free energy expression leads to an even wider range of types of (constant pressure) phase diagrams.

For notational simplicity and for greater transparency of analysis, the χ_{SANS} parameter of Eq. 10 is rewritten in the more compact form

$$\frac{\chi_{SANS}}{C} = a + \frac{b + c\varphi}{T}, \tag{11a}$$

where

$$a = \left(\frac{r_1 - r_2}{z}\right)^2, \tag{11b}$$

$$b = \frac{\varepsilon}{k}\left[\frac{z-2}{2} + \frac{-2p_1 + p_2}{z}\right], \tag{11c}$$

and

$$c = \frac{\varepsilon}{k}\frac{3(p_1 - p_2)}{z}. \tag{11d}$$

Equation 11a emphasizes the presence of an entropic component a and a composition-dependent energetic term, proportional to c, that both vanish when the two blend components have identical monomer structures (and hence $r_1 = r_2$ and $p_1 = p_2$). The sign of b determines whether the phase separation may be of upper critical solution temperature (UCST) or lower critical solution temperature (LCST) types. The Flory formula for χ in Eq. 1 admits only of UCST phase diagrams or complete miscibility, but the SLCT predicts [47, 48] the possibility of an LCST phase separation when the entropic term a is sufficiently large and the exchange energy ε is negative. Notice that this mechanism for the occurrence of LCST phase behavior for binary blends departs from the customary explanation [9, 61] for LCST phase diagrams in polymer solutions: When the temperature of a polymer solution increases, the densities of the pure polymer and pure solvent become increasingly disparate, producing an entropic penalty toward mixing and, therefore, an LCST phase diagram. The LCST of polymer solutions generally appears at a temperature about 70–90% of the critical temperature of the solvent. Hence, the solvent is near the boiling point, while the pure polymer is still either solid or liquid. On the other hand, the components of binary polyolefin blends, for instance, have very similar densities and coefficients of thermal expansion. Consequently, an LCST phase behavior in these systems cannot be governed by the same mechanism that is applicable to polymer solutions. Rather, the competition between a positive entropic portion of the χ parameter and a negative energetic part of χ appears to drive the LCST phase behavior of these polyolefin blends. Recall that the SLCT assumes the blend to be incompressible. Hence, the appearance of LCST phase behavior for binary blends does not automatically imply that it arises because of compressibility effects.

4.1 Basic Critical Properties of Binary Blends in the SLCT

The critical composition φ_c and the critical temperature T_c are obtained as the solutions to the pair of equations

$$\left.\frac{(\partial^2 \Delta F_{mix})}{\partial \varphi^2}\right|_{T,V} = 0 \tag{12a}$$

and

$$\left.\frac{(\partial^3 \Delta F_{mix})}{\partial \varphi^3}\right|_{T,V} = 0 \,. \tag{12b}$$

The condition of Eq. 12b in conjunction with Eq. 9a produces [47] a quadratic equation for φ_c:

$$\begin{aligned} &- 2ac(1 - \varphi_c)\varphi_c^2 \\ &+ [2c(\lambda - 1)\varphi_c^3 - \{b(\lambda - 1) - c(4\lambda - 1)\}\varphi_c^2 + 2(c - b)\lambda\varphi_c + b]/(M\lambda) = 0 \,, \end{aligned} \tag{13}$$

where $\lambda = M_2/M_1$ is a measure of the polymer size asymmetry. The parameter λ replaces the polymerization index ratio $\lambda_N = N_2/N_1$ of FH theory, where N_i denotes the polymerization index for species i. Specializing the constraint of Eq. 12a to $T = T_c$ and $\varphi = \varphi_c$ implies that the critical temperature T_c depends on φ_c through

$$T_c = \frac{2(b + c\varphi_c)}{\frac{1}{M\varphi_c} + \frac{1}{M\lambda(1-\varphi_c)} - 2a} \,. \tag{14}$$

Since $|c/b|$ is typically small, the largest contribution to the shift of T_c from its FH value arises according to Eq. 14 from a nonzero value of parameter a. An increase of a generally leads to decreased blend miscibility.

The correlation length amplitude ξ is another characteristic property of polymer blends that is strongly influenced by monomer size and shape anisotropy. Within a mean-field approximation, such as the SLCT, which is valid [62–64] sufficiently far from the critical point, the correlation length amplitude ξ_0 is independent of temperature and is defined as $\xi \equiv \xi_0(|T - T_c|/T)^{1/2}$, where ξ designates the static correlation length associated with composition fluctuations [65]. Since a mean-field sum rule exhibits ξ^2 as proportional to the structure factor $S(0)$ in the long wavelength limit, ξ_0 controls both the amplitude $[S(0)]$ and the length scale (ξ) of composition fluctuations. The correlation length amplitude at the critical composition φ_c is related to the SLCT quantities b and c of Eqs. 11c and 11d by

$$\xi_0 = \left[\frac{d_0 T_c}{2|b + c|}\right]^{1/2} , \tag{15}$$

where the square gradient coefficient d_0 can be determined within the random phase approximation [62] (RPA) in terms of the critical composition φ_c and the Kuhn lengths a_1 and a_2 as

$$d_0 = \frac{1}{18}\left[\frac{a_1^2}{s_1\varphi_c} + \frac{a_2^2}{s_2(1-\varphi_c)}\right], \tag{15a}$$

with s_1 and s_2 denoting the numbers of united atom groups in single monomers of blend components 1 and 2, respectively.

The monomer shape and size asymmetry also affects the ranges of T over which mean-field and Ising-type critical behaviors are observed. These temperature ranges are expressed in terms of the Ginzburg number Gi, which provides a rough estimate of the magnitude of the reduced temperature $\tau \equiv (T - T_c)/T$ at which the crossover from mean-field to Ising-type behavior occurs. We have introduced [65] a more refined criterion for specifying the three different regimes. Mean-field theory holds for $\tau \geq 10\,\mathrm{Gi}$, while the Ising critical behavior corresponds to $\tau \leq \mathrm{Gi}/10$. The range $\mathrm{Gi}/10 < \tau < 10\,\mathrm{Gi}$ describes a crossover domain with $\tau \approx \mathrm{Gi}$ in the middle of this range. For an incompressible blend, Gi is given by

$$\mathrm{Gi} = \frac{v_{\mathrm{cell}}^2\left[M^{-1}\varphi_c^{-3} + (M\lambda)^{-1}(1-\varphi_c)^{-3}\right]^2}{64\pi^2\left|(M\varphi_c)^{-1} + (M\lambda)^{-1}(1-\varphi_c)^{-1} - 2ad_0^3\right|}, \tag{15b}$$

where v_{cell}^2 is the volume associated with a single united atom group and a and d_0 are defined by Eqs. 11b and 15a, respectively. The article by Schwahn in this volume discusses the details of the crossover analysis required to extract from neutron scattering data quantities suitable for comparison with mean-field theories. Schwahn's review also demonstrates the importance of strong thermal composition fluctuations in modifying blend thermodynamic properties in the vicinity of the critical point.

In analogy to the description of polymer solutions where the theta temperature T_θ is normally identified as an essential reference temperature [4, 62, 66], T_θ can also be defined for dilute polymer blends. Since either of the two components may be the dilute species, there are two osmotic virial expansions and two theta temperatures whose SLCT expressions are [47]

$$T_\theta^{(1)} = \frac{2(b+c)M_1}{1-2aM_1} \tag{16a}$$

and

$$T_\theta^{(2)} = \frac{2bM_2}{1-2aM_2}. \tag{16b}$$

The reduced temperature gap δT_θ between the theta and critical temperatures, $\delta T_\theta^{(i)} \equiv (T_\theta^{(i)} - T_c)/T_c$, is another useful property for characterizing blends. When $M_1 = M_2$, both theta temperatures coincide $T_\theta^{(1)} \approx T_\theta^{(2)} \equiv T_\theta$, and $\delta T_\theta^{(1)} \approx \delta T_\theta^{(2)}$ since $|c/b|$ is generally small [47, 48].

Based on an analysis of the critical parameters (φ_c, T_c, ξ_0, Gi, $\delta T_\theta^{(1)}$, $\delta T_\theta^{(2)}$) determined from Eqs. 13–15, binary polymer blends may be classified into four distinct classes of critical behavior.

4.2 Essential Characteristics of the Four Classes of Binary Blend Miscibility

This subsection provides a brief description of the four classes of binary blend miscibility patterns that are predicted by the SLCT. Eight potential types of binary blend phase behavior can be derived form the SLCT because b may be positive or negative and because a and c may each be either zero or nonzero. However, only four blend categories are classified since the remaining four types of blends are either completely miscible or completely immiscible systems. (Note that Eq. 11b implies that a is nonnegative in the SLCT. Classes of critical behavior for $a < 0$ are not discussed here as they do not emerge from the SLCT.)

Class I corresponds to blends specified by $a = 0$, $b > 0$, and either $c = 0$ or $c \neq 0$. Blends in this class exhibit UCST phase diagrams that are very similar to those predicted by FH theory. The basic quantities scale with $M \equiv M_1$ as

$$\begin{aligned}
&\varphi_c \approx \varphi_c^{(I)} \equiv \frac{\lambda^{1/2}}{1+\lambda^{1/2}}\,,\\
&T_c \sim M\,,\\
&\xi_0 \sim M^{1/2}\,,\\
&\mathrm{Gi} \sim M^{-1}\,,\\
&\delta T_\theta^{(1)} \sim \frac{1+2\lambda^{1/2}}{\lambda}\,, \quad \delta T_\theta^{(2)} \sim 1+2\lambda^{1/2}\,,
\end{aligned} \tag{17.I}$$

where $\lambda = M_2/M_1$ and $\varphi_c^{(I)}$ is a generalization of the FH theory critical composition $\varphi_c^{(FH)} = \lambda_N^{1/2}/[1+\lambda_N^{1/2}]$ to structured monomer blends.

The three other classes of blends display qualitative departures from the FH type behavior of class I. Class II is defined by $a \neq 0$, $b > 0$, and either $c \neq 0$ or $c = 0$ and is also characterized by a UCST phase diagram, but a nonlinear dependence of T_c on M. The critical properties vary with M as

$$\begin{aligned}
&\varphi_c \approx \varphi_c^{(I)}\,,\\
&T_c \sim \frac{M}{1-2alM}\,,\\
&\xi_0 \sim \{M/[1-2alM]\}^{1/2}\,,
\end{aligned}$$

$$
\begin{aligned}
&\mathrm{Gi} \sim \frac{M^{-1}}{1-2alM}, \\
&\delta T_\theta^{(1)} \sim \frac{1+2\lambda^{1/2}}{\lambda(1-2aM)}, \quad \delta T_\theta^{(2)} \sim \frac{1+2\lambda^{1/2}}{(1-2aM)},
\end{aligned} \tag{17.II}
$$

where $l \equiv \lambda/[1+\lambda^{1/2}]^2$. The occurrence of the class II type miscibility pattern is limited to $M < M^* = 1/2al$ because the expression for T_c diverges for M^*.

Class III and IV type blends both have LCST miscibility patterns. Class III corresponds to $a \neq 0$, $b < 0$, and $c = 0$, and the predicted T_c for this class no longer scales as a power of M but instead approaches the constant $|b|/a$ as $M \to \infty$:

$$
\begin{aligned}
&\varphi_c = \varphi_c^{(I)}, \\
&T_c = |b|/a, \\
&\xi_0 = \text{constant}, \\
&\mathrm{Gi} \sim M^{-2}, \\
&\delta T_\theta^{(1)} \sim M^{-1}, \quad \delta T_\theta^{(2)} \sim M^{-1}.
\end{aligned} \tag{17.III}
$$

Class IV ($a \neq 0$, $b < 0$, and $c \neq 0$) exhibits the strongest departures from the predictions of FH theory since the dependence of both T_c and φ_c on M deviates significantly from those emerging from FH theory. The critical properties of class IV blends approach different limits depending on the sign of c. The scaling laws in the $M \to \infty$ limit for $c < 0$ are

$$
\begin{aligned}
&\varphi_c \sim M^{-1/2}, \\
&T_c = |b|/a, \\
&\xi_0 \sim M^{1/4}, \\
&\mathrm{Gi} \sim M^{-1/2}, \\
&\delta T_\theta^{(1)} = c/b, \\
&\delta T_\theta^{(2)} \sim M^{-1/2}.
\end{aligned} \tag{17.IVa}
$$

The sign of c is reversed upon interchanging the labels of the two components, but the scaling laws for $c > 0$ are not symmetric to those in Eq. 17.IVa, as evident from the following:

$$
\begin{aligned}
&\varphi_c = 1 - O\left(M^{-1/2}\right), \\
&T_c = |b-c|/a, \\
&\xi_0 \sim M^{1/4}, \\
&\mathrm{Gi} \sim M^{-1/2}, \\
&\delta T_\theta^{(1)} \sim M^{-1/2}, \\
&\delta T_\theta^{(2)} = c/|b-c|.
\end{aligned} \tag{17.IVb}
$$

Figure 4a–e illustrates the dependence of the critical properties for some *symmetric* polyolefin blends ($M_1 = M_2 = M$; $\lambda = 1$) specified in the figure caption. The linear scaling of T_c with M in class I is one hallmark of FH theory [3–7]; blends of class II exhibit a stronger than linear dependence of T_c on M; while the critical temperatures for those in classes III and IV approach constants as $M \to \infty$. The critical composition φ_c is insensitive to M for symmetric blends of classes I–III but for class IV decreases toward zero (or unity) with the scaling $\varphi_c \sim M^{-1/2}$. Hence, the FH estimate for φ_c can be grossly in error for LCST blends. The correlation length amplitude ξ_0 for blends of class II increases more rapidly with M than for class I ($\xi_0 \sim M^{1/2}$), but ξ_0 for blends of classes III and IV is significantly smaller than the chain radius of gyration, $R_g \sim M^{1/2}$. An insensitivity of ξ_0 to M occurs for class III blends where ξ_0 is comparable to the statistical segment length rather than to R_g. The theta point shift δT_θ is also a useful indicator of the blend miscibility class. The

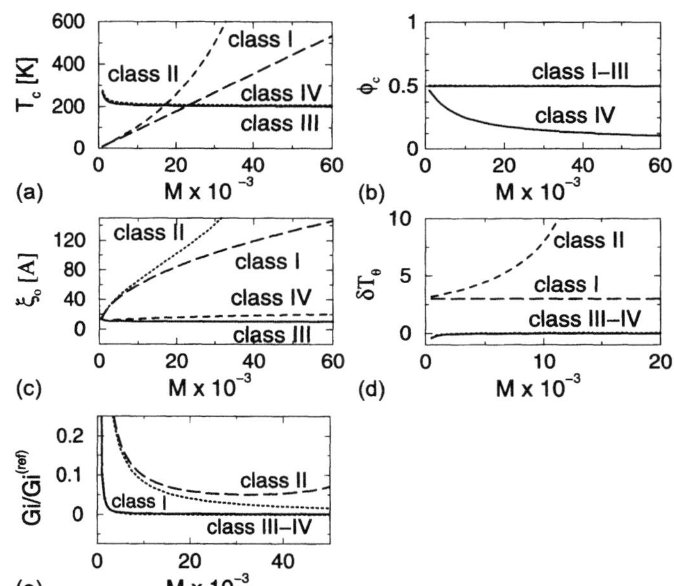

Fig. 4 The critical temperature T_c (**a**) critical composition φ_c (**b**) correlation length amplitude ξ_0 (**c**) reduced theta temperature $\delta T_\theta = (T_\theta - T_c)/T_c$ (**d**) and the reduced Ginzburg number $\mathrm{Gi}/\mathrm{Gi}_i^{(\mathrm{ref})}$ (**e**) for symmetric blends ($\lambda = 1$, $M_1 = M_2 = M$) as a function of the number of united atom groups M in a single chain. ($\mathrm{Gi}_i^{(\mathrm{ref})} = 0.01$ is a typical value of Gi for mixtures of small molecules.) Classes II and IV are represented by PH1/PEP and PIB/PEP blends, respectively. The exchange energy ε for the PH1/PEP and PIB/PEP systems are taken as $\varepsilon = 0.01$ K and $\varepsilon =- 1$ K, respectively. The example for class I blends ($a = c = 0$, $b > 0$) is constructed assuming the same b as is chosen for class II, while the example for class III blends ($a \neq 0$, $c = 0$, $b < 0$) is generated by taking a and b equal to those for the PH1/PEP mixture

large values of $\delta T_\theta \equiv \delta T_\theta^{(1)} \approx \delta T_\theta^{(2)}$ for blends of classes I and II suggest that the theta point is experimentally inaccessible. (Most polymers with T_c near room temperature would become thermally unstable at these T_θ.) In contrast, T_θ is predicted to lie close to T_c for blends of classes III and IV.

The scaling of Gi for FH type (class I) blends [62–64, 67], Gi $\sim M^{-1}$, indicates a strong decrease of the width of the critical region with increasing M. However, the magnitude of Gi can be larger for class II blends that exhibit a shallow *minimum* in Gi as a function of M (Fig. 4e). A typical magnitude for the minimum value of Gi for class II blends is about 0.001, which is small relative to typical values for Gi of small-molecule mixtures (Gi ≈ 0.01) [68, 69]. The Gi for class III blends decreases even more rapidly than Gi for class I, while the dependence of Gi on M for class IV (Gi $\sim M^{-1/2}$) is weaker than for class I, resembling the Gi scaling for UCST polymer solutions [70]. Apparently, the examples of class II–IV blends indicate the presence of large departures from the Gi $\sim M^{-1}$ scaling of FH theory, but there is nevertheless a general tendency for Gi to become small for large M.

4.3
Experimental Support for New Classes of Polymer Blend Miscibility

A linear scaling of T_c with M has been confirmed by SANS experiments [71] for symmetric ($N_1 = N_2$) isotopic polyolefin blends (where monomer molecular structures are almost identical) and by Monte Carlo simulations [72, 73] for symmetric linear chain polymer blends. Several experiments, on the other hand, indicate that this FH pattern of blend miscibility is not general. Perhaps the best documented example of a gross departure from FH type phase behavior is provided by the polystyrene/poly(vinyl methyl ether) (PS/PVME) blend whose T_c is found to be nearly insensitive to M ($T_c = 145 \pm 5\,°\mathrm{C}$), while φ_c is observed [74, 75] to be highly asymmetric even for samples with identical N_1 and N_2 [74, 75]. Based on our fits [17] to the scattering data that yield $b < 0$, $a > 0$, and $c \neq 0$, PS/PVME systems are classified as class IV blends for which the SLCT predicts that T_c approaches the ratio $|b|/a$ as $M \to \infty$, a prediction that is in accord with the observed insensitivity of T_c to variations in M. In addition, the SLCT scaling relation $\varphi_c \sim M^{-1/2}$ is consistent with the finding of Han and coworkers [74, 75] that increasing M by a factor of 3 reduces φ_c by roughly a factor of 2. Other SLCT predictions that the correlation length amplitude ξ_0 for PS/PVME blends weakly depends on M and that T_c and T_θ are almost equal also agree with the existing experimental data [74–77].

A similar insensitivity of T_c to M has been observed [60] for binary blends of poly(isobutylene) (PIB) with several other polyolefins, systems for which it is difficult to imagine that any *specific interactions* could be responsible for the observed dramatic departures from the predictions of FH theory. (Note that our explanation for the unusual critical behavior of PS/PVME blends likewise does not resort to invoking the presence of *specific interactions*.) The LCST phase separation in binary blends of PIB with other polyolefins arises according to the SLCT from the competition between a negative energetic portion of the χ_{SANS} parameter and a sufficiently positive *entropic* portion a of χ_{SANS}. The large values of a for the PIB/polyolefin blends stem from a rather large partial entropic parameter $r_{PIB} = 7/4$, which, in turn, arises because of the presence of the tetrafunctional carbon atom in each PIB monomer (Eq. 8). A negative exchange energy ε (implying a negative b) may occur because 50% of the PIB united atom groups are CH_3 groups that have larger attractive interactions (i.e., Lennard–Jones potential interaction parameters) than the CH_2, CH, and C united atom groups [37–39]. This effect produces a larger self-interaction energy $\varepsilon_{11} = \varepsilon_{PIB-PIB}$ (relative to ε_{22}) and a larger heterocontact interaction energy ε_{12} with other polyolefins, leading to a negative $\varepsilon = \varepsilon_{11} + \varepsilon_{22} - 2\varepsilon_{12}$.

5
Application of the SLCT to LCST Polyolefin Blends

LCST polyolefin blends are quite unusual systems because the dispersive van der Waals nature of the interactions between polyolefin species favors UCST phase separation, as found in the majority of binary polyolefin mixtures. LCST phase diagrams have been observed [60] in binary blends of poly(isobutylene) (PIB) with, for instance, poly(propylene) (PP), head-to-head poly(propylene) (hhPP), saturated polybutadienes (sPB), or poly(ethylene propylene) (PEP). All these LCST systems exhibit large negative values of the χ_{SANS} parameter at lower temperatures and positive χ_{SANS} at higher temperatures [60]. Not surprisingly, this LCST behavior cannot be explained by solubility parameter theory, which has been applied with apparent success to a number of UCST polyolefin blends by Graessley et al. [27–32]. Thus, the thermodynamic description of these LCST blends of PIB presents a significant challenge to any polymer theory concerned with modeling the molecular factors controlling blend miscibility. Consequently, these blends of PIB with other polyolefins are analyzed here using the high pressure, high molecular weight limit of the LCT.

The incompressibility assumption implies that the SLCT for binary blends contains only one adjustable parameter, the exchange energy ε. All remaining quantities of the SLCT (i.e., r_α and p_α, $\alpha = 1, 2$) are determined from the united atom group monomer structures (Fig. 1a) by elementary counting. Figure 5 compares the calculated and experimental χ_{SANS} parameters for poly(isobutylene)/head-to-head polypropylene (PIB/DhhPP) blends as a function of the inverse temperature. Squares denote the experimental data

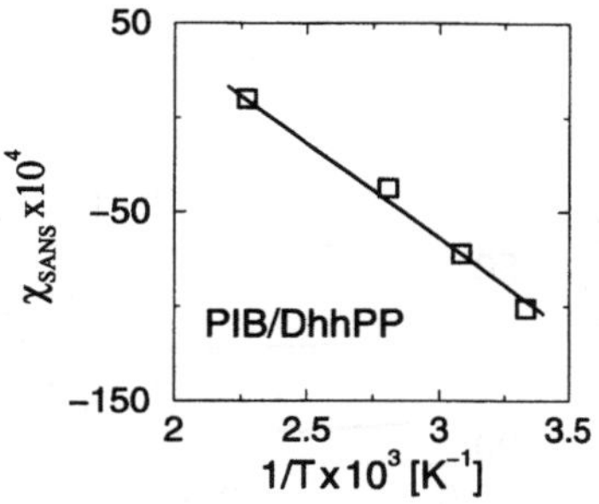

Fig. 5 Comparison of calculated and experimental interaction parameter χ_{SANS} for PIB/DhhPP blends ($\varphi_{PIB} = 0.475$) as a function of inverse temperature. *Squares* are the experimental data of Krishnamoorti et al. [60], while the *solid line* is a least-squares fit of the SLCT χ_{SANS} for the exchange energy $\varepsilon = -1.16$ K

of Krishnamoorti et al. [60] for $\varphi_{PIB} = 0.475$, while the solid line represents a least-squares fit of the SLCT Eq. 11a to the data for χ_{SANS}. The least-squares fit yields $\varepsilon = -1.16$ K, guaranteeing that the theory reproduces the slope of χ_{SANS} vs. $1/T$, but the remarkable feature of Fig. 5 is the excellent theoretical representation for the entropic portion of χ_{SANS}.

Given the empirical value for ε, the SLCT relation in Eq. 10 can then be used to generate predictions for the composition dependence of the χ_{SANS} parameter, which to our knowledge has not yet been studied. This composition dependence is delineated for four different temperatures in Fig. 6. The squares correspond to the experimental data [60] that are available only for the single composition $\varphi_{PIB} = 0.475$. The convex parabolic variation of $\chi(\varphi)$ for PIB blends in Fig. 6 becomes less pronounced at higher temperatures, almost disappearing for $T \geq 440$ K.

Figure 7 further illustrates the influence of monomer molecular structures on the miscibility of PIB blends by presenting computed spinodal curves for three binary blends of PIB with other polyolefins. The *homologous* nature of the second component provides a motivation for using the same PIB/DhhPP exchange energy $\varepsilon = -1.16$ K in the SLCT calculations for all three systems. The computed phase diagrams (in Fig. 7) accord with the experimental observations that the PIB/DhhPP blend is a strongly miscible system, whereas mixtures of PIB with either poly(ethyl propylene) (PEP) or saturated 1,2-polybutadiene (H1,2PB) are phase separated at room temperatures. The computed critical temperature T_c for the PIB/H1,2PB blend is higher by at least 20 K than the T_c deduced from the visual observation [60] of cloud points for this system, a small difference considering that the same exchange energy ε has been assumed for PIB/H1,2PB and PIB/DhhPP mixtures. Experiments for a 50–50 mixture of PIB and DPEP reveal [60] that this system is phase separated at 298 K, whereas the computed spinodal temperature is 260 K (Fig. 7), consistent with this observation.

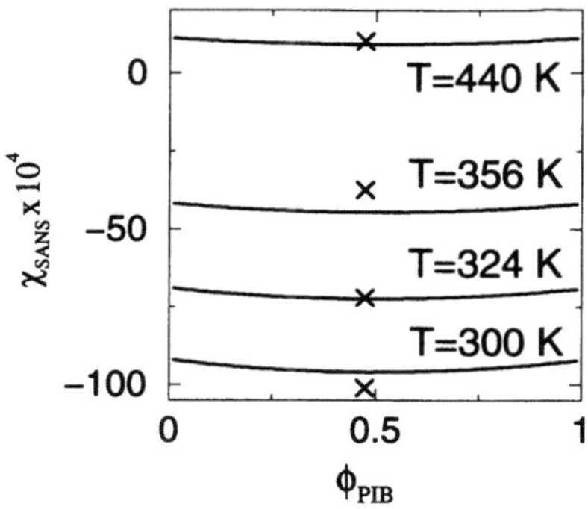

Fig. 6 SLCT predictions for the composition dependence of the interaction parameter χ_{SANS} for PIB/DhhPP blends ($\varphi_{PIB} = 0.475$) at several temperatures. *Crosses* represent experimental data of Krishnamoorti et al. [60]. The exchange energy $\varepsilon = -1.16$ K is taken from the fit in Fig. 5

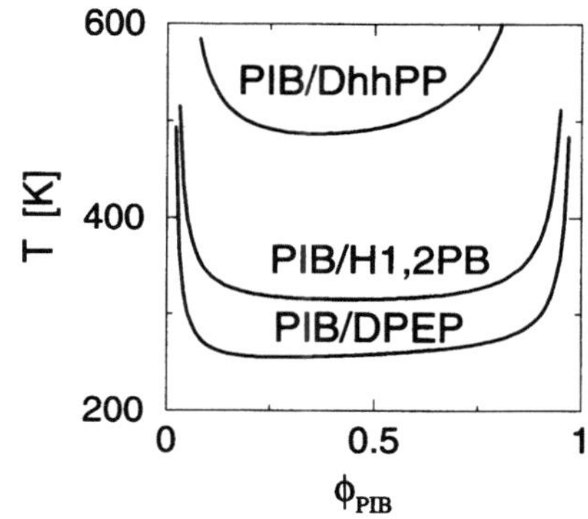

Fig. 7 SLCT spinodal curves for PIB/DhhPP, PIB/H1,2PB, and PIB/DPEP blends that are computed by assuming the common exchange energy $\varepsilon = -1.16$ K for all three systems. The molecular weights of the blend components correspond to the experimental samples studied by Krishnamoorti et al. [60]

6
Introduction of Chain Stiffness into the SLCT

The model employed by the SLCT so far treats each polymer as a completely flexible entity, subject only to the excluded volume constraints against multiple occupancy of a lattice site by any two united atom groups. However, bond angle constraints and steric interactions of, e.g., hydrogen atoms and side groups lead to local chain stiffness and the semiflexibility of polymer molecules. For example, torsional motions about single C – C backbone bonds in many olefins are limited to a *trans* and a pair of *gauche* conformations with differing energies and, therefore, temperature-dependent probabilities of occurrence. Following Flory [78], these different conformational states are modeled on a simple cubic lattice by defining the trans conformation as the collinear configuration of two consecutive bonds, i.e., the two bonds lie along the same direction, and by identifying the gauche conformation with situations in which the two consecutive bonds reside along orthogonal directions [79]. The energy of the trans configuration is taken as zero, while that of the gauche configuration is $E_b^{(i)}$, where the superscript i denotes the chain species.

This definition of trans and gauche configurations is quite clear for linear chains of polyethylene, but requires modification when the monomers contain side groups (Fig. 1). For example, the methyl side groups in poly(propylene) must lie in a direction orthogonal to at least one of the backbone bonds that emanate from a backbone trifunctional CH group. Thus, there is no physical sense in assigning a gauche energy penalty to this conformation. Hence, only the backbone bonds in poly(propylene) can have the bending energies $E_b^{(i)}$ in the LCT, and, consequently, the maximum number of gauche conformations (ignoring excluded volume constraints) in a poly(propylene) chain is $N_b - 2$, where N_b is the number of backbone C – C bonds. Further complexity arises for polymer species in which each side group contains two or more united atom units because a pair of successive side group bonds may have either a trans or gauche configuration. Consequently, the maximum number of gauche conformations (again ignoring excluded volume constraints) in a poly(ethyl ethylene) chain is $N_b - 2 + (1/2)N_b$, where the extra $(1/2)N_b$ contribution stems from the presence of the side groups. Although it is possible to assign different gauche energies $E_b^{(i)}$ to pairs of backbone or side group bonds [49], a single energy $E_b^{(i)}$ is assumed for each species in order to minimize the number of adjustable parameters. The *bending* energies $E_b^{(i)}$ are determined by fitting calculated thermodynamic properties to experimental data. Because of the inadequacy of a cubic lattice to represent the actual numbers and geometrical structures of the trans and gauche configurations in real olefins, it is best not to determine the Flory model *bending* energies $E_b^{(i)}$ from fits to structural data such as the average chain radii of gyration. The Flory *trans–gauche* model [78], with one trans and $(z - 2)$ gauche conformations for a bond pair, thus emerges as a simple method for including the thermodynamic consequences of local chain stiffness.

The introduction of chain semiflexibility into the SLCT leads to the replacement [46, 49, 80, 81] of the geometrical parameters r_i and p_i in Eqs. 9a, 10, and 11b–11d by polynomials in the bending energy statistical factor

$$g_i = \frac{z}{z - 1 + \exp\left[E_b^{(i)}/kT\right]}, \tag{18}$$

where z is the lattice coordination number, which equals 6 for a simple cubic lattice. The resultant polynomials have the forms

$$r_i \to r_i^{(\mathrm{semi})}(g_i) = \alpha_0^{(i)} + \alpha_1^{(i)} g_i \,, \tag{19a}$$

$$p_i \to p_i^{(\mathrm{semi})}(g_i) = \gamma_0^{(i)} + \gamma_1^{(i)} g_i + \gamma_2^{(i)} g_i^2 \,, \tag{19b}$$

where the coefficients $\alpha_0^{(i)}$, $\alpha_1^{(i)}$, $\gamma_0^{(i)}$, $\gamma_1^{(i)}$, and $\gamma_2^{(i)}$ are partitions of the SLCT parameters r_i and p_i,

$$\alpha_0^{(i)} + \alpha_1^{(i)} = r_i \,, \tag{19c}$$

$$\gamma_0^{(i)} + \gamma_1^{(i)} + \gamma_2^{(i)} = p_i \,. \tag{19d}$$

The fully flexible chain limit is recovered from Eqs. 19a and 19b by setting $E_\mathrm{b}^{(i)} = 0$ or, equivalently, $g_i = 1$:

$$r_i^{(\mathrm{semi})}(g_i = 1) = \alpha_0^{(i)} + \alpha_1^{(i)} = r_i \,,$$

$$p_i^{(\mathrm{semi})}(g_i = 1) = \gamma_0^{(i)} + \gamma_1^{(i)} + \gamma_2^{(i)} = p_i \,.$$

Due to the existence of Euler relations for $N_2^{(i)}$, the coefficients $\alpha_0^{(i)}$ and $\alpha_1^{(i)}$ can simply be related [46] to the numbers $s_\mathrm{i}^{(\mathrm{tri})}$ and $s_\mathrm{i}^{(\mathrm{tetra})}$ of tri- and tetra-functional united atom groups in a single monomer of species i as

$$\alpha_0^{(i)} = 2\left[\frac{s_\mathrm{i}^{(\mathrm{tri})}}{s_i} + 2\frac{s_\mathrm{i}^{(\mathrm{tetra})}}{s_i}\right]$$

and

$$\alpha_1^{(i)} = 1 - \frac{s_\mathrm{i}^{(\mathrm{tri})}}{s_i} - 3\frac{s_\mathrm{i}^{(\mathrm{tetra})}}{s_i} \,.$$

The absence of a Euler relation for $N_3^{(i)}$ for arbitrary monomer structures renders the calculation of $\gamma_0^{(i)}$, $\gamma_1^{(i)}$, and $\gamma_2^{(i)}$ much more involved. The coefficients $\gamma_0^{(i)}$, $\gamma_1^{(i)}$, and $\gamma_2^{(i)}$ are obtained by counting the numbers $n_j^{(i)}$ of configurations for three successive bonds that traverse a monomer of species i and that have $j = 0$-, 1-, or 2-bond pairs belonging to identical subchains. More specifically,

$$\gamma_j^{(i)} = \frac{n_j^{(i)}}{s_i} \,, \quad j = 0, 1, 2 \,.$$

The concept of subchains is explained in several prior papers [49, 80, 81], which may be consulted for more details. Briefly, the chain backbone is the first subchain. Other subchains are formed from sequences of side group bonds to which gauche conformations may be assigned.

Because the introduction of chain semiflexibility imparts a temperature dependence to $r_i^{(\mathrm{semi})}$ and $p_i^{(\mathrm{semi})}$, the free energy of mixing ΔF_mix and the χ_SANS parameter vary with temperature in a more complicated fashion than these quantities vary in the completely flexible chain SLCT. Equations 13 and 14 for the critical temperature and composition of a binary, incompressible blend are still maintained in the semiflexible chain extension of the SLCT, but now a, b, and c are functions of temperature. Consequently, the pair of equations for T_c and φ_c must be solved numerically. While it is expected that the semiflexible chain SLCT admits of a wider range of miscibility patterns than those discussed in Sect. 4.2, a detailed analysis has yet to be made.

It is straightforward to reformulate the SLCT in order to account for rigidity within any of the components of the blend because a pair of rigid bonds corresponds to a situation in which the two semiflexible bonds have an infinite bending energy $E_b^{(i)}$ or, equivalently, a vanishing g_i. Thus, the rigid bond constraint implies that these rigid bond configurations are effectively excluded from the set of configurations contributing to the geometrical parameters r_i and p_i or $r_i^{(\mathrm{semi})}$ and $p_i^{(\mathrm{semi})}$. Section 8.2.1 describes the evaluation of these geometrical parameters for norbornene-co-ethylene binary blends where the rigidity of the norbornene monomers is modeled by taking the side group bonds in the structure of Fig. 10b as completely rigid.

7
Application of the SLCT to UCST Polyolefin Blends

The majority of binary polyolefin blends exhibit upper critical solution temperature (UCST) phase diagrams. The earlier common view that these UCST polyolefin mixtures are not of great scientific interest has been dispelled due to recent experiments by Graessley et al. [31] that reveal quite different miscibilities of poly(propylene) (PP) and head-to-head poly(propylene) (hhPP) in binary blends with poly(ethylene propylene) (PEP). The critical temperatures differ by more than 100 K for blends with polymerization indices in the hundreds, as illustrated by the LCT spinodal curves [82] in Fig. 8 that have been computed for binary mixtures of either PP or hhPP with PEP. This huge difference in miscibility arises merely from the *swapping* of alternate methyl side groups along the poly(propylene) chain.

While the spinodal curves [82] depicted in Fig. 8 refer to LCT calculations for compressible blends of semiflexible polymers at a pressure of 1 atm, similar behavior emerges [83] from the SLCT where the pressure is infinite. Thus, we analyze the molecular features contributing to this remarkable difference in miscibilities within the framework of the SLCT because of its analytical tractability. In order to elucidate the contributing features to the observed miscibility difference, it is convenient to assume first that both blend components are completely flexible. Then, the partial entropic structural parameters r_{PP} and r_{hhPP} are identical (Sect. 3.1), so that all of the calculated miscibility differences between PP/PEP and hhPP/PEP blends emerge from the differ-

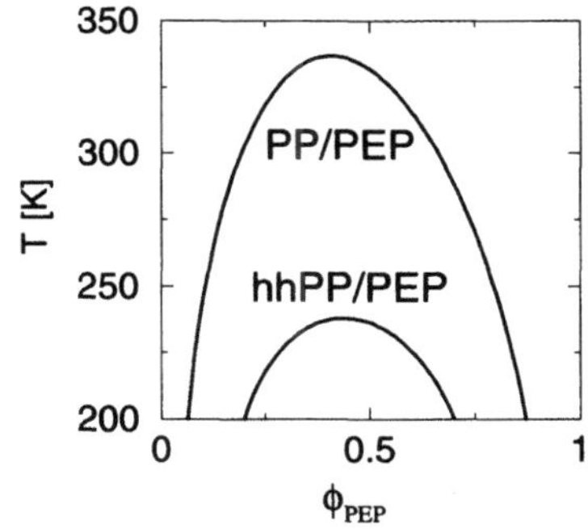

Fig. 8 LCT spinodal curves for PP/PEP and hhPP/PEP blends at $P = 1$ atm. The site occupancy indices $M_{\text{PP}} = 1560$ and $M_{\text{PEP}} = 4275$ used in the calculations correspond to one of the blends studied by Graessley et al. [31], and $M_{\text{hhPP}} = 1560$ is selected to ensure the same molecular weights for the PP and hhPP components. Both blends are assumed to have the same interaction energies $\{\varepsilon_{ij}\}$ in order to illustrate the sole influence of monomer structure and chain semiflexibility on the blend miscibilities. The self-interaction energies $\varepsilon_{11} = (1/2)(\varepsilon_{\text{PP-PP}} + \varepsilon_{\text{hhPP-hhPP}}) = 205.40$ K and $\varepsilon_{22} = \varepsilon_{\text{PEP-PEP}} = 207.49$ K are obtained from our earlier fits to PVT data for the pure melts. The bending energies $E_{\text{b}}^{(\text{PP})} = 219$ K, $E_{\text{b}}^{(\text{hhPP})} = 277$ K, and $E_{\text{b}}^{(\text{PEP})} = 460$ K are taken form [80], while the heterocontact energy is an adjustable parameter

ent geometrical indices p_{PP} and $p_{hh\text{PP}}$. Inspection of Fig. 1b indicates that the united atom structure for hhPP generates one three-bond run not present for the PP structure, namely, the three-bond run that comprises the bonds between two adjacent methyl side groups and the chain backbone along with the joining backbone bond. Thus, use of Eq. 10 implies that hhPP presents a lower surface fraction for interaction with PEP than does PP, i.e., hhPP has a lower total probability of heterocontact interactions with PEP than PP, leading to a higher compatibility of hhPP/PEP blends compared to PP/PEP mixtures. (Because molecular weights have enormous influence on the critical temperatures, the calculated spinodal curves in Fig. 8 have been determined assuming common molecular weights for PP and hhPP and the same molecular weight for PEP in both blends.)

Additional contributions to the disparate miscibilities of hhPP/PEP and PP/PEP blends arise from the obvious difference in chain stiffness between PP and hhPP that is induced by the presence of methyl side groups on adjacent backbone carbons in hhPP. As explained below, lifting the assumption that the polymer chains are fully flexible provides additional factors favoring the miscibility of PP/PEP blends with respect to PP blends. Clearly, the presence of methyl groups attached to adjacent backbone carbons in hhPP leads to greater steric interactions in hhPP than in PP. Thus, a larger trans $\leftrightarrow$ gauche energy difference $E_{\text{b}}^{(i)}$ must be ascribed to hhPP, which in turn implies the presence of a nonzero difference in the entropic χ_{SANS} between hhPP/PEP and PP/PEP blends. This stiffness-induced entropic χ_{SANS} favors the miscibility of the hhPP/PEP blends. In other words, the greater stiffness of hhPP enables improved packing of hhPP with PEP than of PP with PEP. In addition, the difference in chain stiffness affects the magnitudes of the geometric indices $p_{PP}^{(semi)}$ and $p_{hhPP}^{(semi)}$ that influence the energetic portion of χ_{SANS}. Again, the greater stiffness of hhPP reduces the probability of heterocontacts with PEP as compared to PP. In summary, the greater miscibility of hhPP/PEP blends over PP/PEP blends stems from the synergistic combination of three factors: an inherently lower surface fraction for hhPP, a stiffness-induced improved packing for hhPP, and a stiffness-driven diminished probability for hhPP–PEP heterocontact interactions [82].

7.1
Analysis of the PP/hhPP blend

Binary blends of PP with hhPP represent another interesting system from a molecular standpoint. Fitting the experimental χ_{SANS} for a symmetric ($\varphi = 0.5$) PP/DhhPP mixture [31] to the common empirical form $\chi = A + B/T$ yields [46] a negative $A = -4.46 \times 10^{-3}$ and a large $B = 2.38$ K (per monomer). The value of B is typical for fairly immiscible UCST blends, in sharp contrast to intuitive expectations that the almost chemical identity of PP and hhPP should lead to a nearly vanishing exchange energy, i.e., a very small B. Moreover, the rather large, negative A appears inconsistent with the SLCT Eq. 9, which defines the entropic portion of χ_{SANS} as a nonnegative quantity. Because differences in steric interactions are anticipated to be crucial factors driving the system's thermodynamics for mixtures of chemically very similar components, both PP and hhPP are treated within the SLCT as semiflexible polymers.

The parameters $r_i^{(semi)}$ and $p_i^{(semi)}$ for PP and hhPP are evaluated as

$$r_{PP}^{(semi)} = \frac{2}{3} + \frac{2}{3} g_{PP}, \tag{20a}$$

$$g_{PP} = \frac{z}{z - 1 + \exp\left[E_b^{(PP)}/kT\right]},$$

$$r_{hhPP}^{(semi)} = \frac{2}{3} + \frac{2}{3} g_{hhP}, \tag{20b}$$

$$g_{hhPP} = \frac{z}{z - 1 + \exp\left[E_b^{(hhPP)} kT\right]},$$

$$p_{PP}^{(semi)} = \frac{2}{3} + \frac{2}{3} g_{PP} + \frac{2}{3} g_{PP}^2, \tag{20c}$$

$$p_{hhPP}^{(semi)} = \frac{1}{6} + \frac{2}{3} g_{hhPP} + \frac{2}{3} g_{hhPP}^2, \tag{20d}$$

and the corresponding SLCT expression for χ_{SANS} is obtained from Eq. 10 by replacing r_i and p_i by $r_i^{(\mathrm{semi})}$ and $p_i^{(\mathrm{semi})}$, respectively:

$$\frac{\chi_{\mathrm{SANS}}}{[s_{\mathrm{PP}}s_{\mathrm{hhPP}}]^{1/2}} = \left(\frac{r_{\mathrm{PP}}^{(\mathrm{semi})} - r_{\mathrm{hhPP}}^{(\mathrm{semi})}}{z}\right)^2 + \left(\frac{\varepsilon}{kT}\right)\left(\frac{z-2}{2} + \frac{1}{z}\left\{p_{\mathrm{PP}}^{(\mathrm{semi})}[1-3(1-\varphi)] + p_{\mathrm{hhPP}}^{(\mathrm{semi})}[1-3\varphi]\right\}\right) + O\left(\left(\frac{\varepsilon}{kT}\right)^2\right), \tag{21}$$

where $s_{\mathrm{PP}} = 3$ and $s_{\mathrm{hhPP}} = 6$ are the numbers of united atom groups in individual monomers of PP and hhPP, respectively.

An examination of Eqs. 20a–20c reveals that the term $\left(r_{\mathrm{PP}}^{(\mathrm{semi})} - r_{\mathrm{hhPP}}^{(\mathrm{semi})}\right)^2$ in Eq. 21 is temperature dependent, so that the observed [31] large negative temperature-independent portion of χ_{SANS} can only arise from an interplay between the ε/kT factor and the $p_{\mathrm{PP}}^{(\mathrm{semi})}$ and $p_{\mathrm{hhPP}}^{(\mathrm{semi})}$ parameters that are temperature dependent through the bending energy factors g_{PP} and g_{hhPP}. While the SLCT for blends of semiflexible polymers contains three adjustable parameters (the exchange energy ε and the two bending energies $E_{\mathrm{b}}^{(1)}$ and $E_{\mathrm{b}}^{(2)}$), calculations for various systems suggest that χ_{SANS} effectively depends on the difference $\Delta E_{\mathrm{b}} = E_{\mathrm{b}}^{(1)} - E_{\mathrm{b}}^{(2)}$ in the bending energies for the two components, thereby effectively diminishing the number of adjustable parameters to two [46].

We find [46] a wide range of parameter pairs that produce good agreement with the experimental data [31], and Fig. 9 illustrates two examples of these

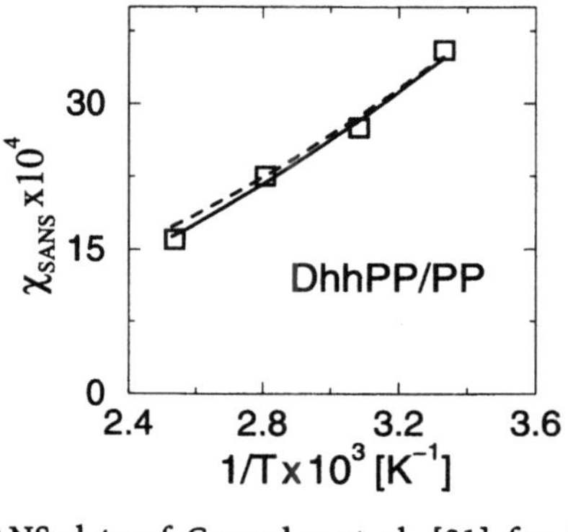

Fig. 9 Two fits to the SANS data of Graessley et al. [31] for the hhPP/PP blend with $M_{\mathrm{hhPP}} = 1920$, $M_{\mathrm{PP}} = 780$, and $\varphi_1 = \varphi_2 = 0.5$. The *solid line* represents one fit obtained for $\varepsilon = 0.000415$ K and $\Delta E_{\mathrm{b}} = E_{\mathrm{b}}^{(\mathrm{hhPP})} - E_{\mathrm{b}}^{(\mathrm{PP})} = 294$ K, while the *dashed line* indicates a fit generated for $\varepsilon = -0.0166$ K and $\Delta E_{\mathrm{b}} = 290$ K

fits for $\varepsilon = -0.0166$ K and $\varepsilon = +0.000415$ K and for $\Delta E_b \equiv E_b^{(hhPP)} - E_b^{(PP)} =$ 290 K and $\Delta E_b = 294$ K, respectively. The minuscule values of the exchange energy ε accord with our expectation that PP and hhPP are almost identical from an interaction energy standpoint. The greater steric interactions between the neighboring side groups in hhPP are likewise anticipated to render hhPP stiffer than PP, in agreement with our fits that yield $E_b^{(hhPP)} > E_b^{(PP)}$.

8
Copolymers Blends

Statistical copolymers are of considerable technological and scientific importance. The earliest theoretical treatments [84–87] of these systems were based on a direct extension [88, 89] of Flory–Huggins theory to random copolymers. While the resultant random copolymer FH theory has been very successful in explaining the enhanced miscibilities [84–86] observed in many systems containing random copolymers, the theory exhibits some serious limitations. First of all, the theory is insensitive to monomer sequence and to the detailed chemical structure of the individual monomers. Some attempts [90–92] to improve the random copolymer FH theory introduce a set of phenomenological interaction parameters $\chi_{ijk,lmn}$ (or $\chi_{ij,kl}$) that explicitly depend on the monomer sequence. However, these phenomenological models are required to invoke additional (often arbitrary) assumptions to reduce the huge number of resulting $\chi_{ijk,lmn}$ parameters to a manageable few. Thus, there is a great need for developing analytically simple, more realistic, improved theories with a minimal number of adjustable parameters. These improved, yet simple, theories should provide an adequate tool for probing general physical trends. An example of this type of simple, molecular-based theory is given by the SLCT approach [46–48] for binary homopolymer blends. A similar, but somewhat more simplified, version of the LCT has also been developed for random copolymer systems. The current section reviews this theory and describes particular applications to blends of random copolymer polyolefins and to diblock copolymer systems.

In addition to the immense technological importance of polyolefins and to the enormous progress [1, 2] in their synthesis that enables the control of their chemical structures, binary blends of polyolefins represent a class of *weakly interacting* systems. Our earlier LCT predictions [22] that slight modifications of chemical structure can significantly alter the properties of blends have been confirmed by the observation of considerably different miscibilities in a variety of polyolefin mixtures [27–32, 60], for example, a large body of experimental data for polyolefin blends [31, 60], the majority of which have one or both of the components as random copolymers. These data indicate that there is a strong sensitivity of the thermodynamic properties of binary polymer blends to the monomer molecular structures of the blend components. An understanding of this relationship can only be achieved within a molecular-based theory for statistical copolymers systems, which are quite prevalent in technological applications.

The development of the equivalent of the homopolymer blend SLCT (reviewed in Sect. 3) for blends containing random copolymers has been extremely lengthy and algebraically rather complicated [93]. This complexity arises not only from the random nature of the copolymers (i.e., from the statistical positioning of the different monomer species along the polymer chain) but also from the fact that local correlations, which are responsible for nonrandom mixing effects, must be averaged over distances that are larger than a few bond lengths in order to provide an adequate description of the dependence of the system's free energy on the monomer sequence [94]. Consequently, the SLCT free energy expressions for copolymer blends does not exhibit [93] the analytical simplicity of the SLCT Eq. 9a for homopolymer blends. Hence, additional assumptions are introduced for copolymer systems beyond the neglect of excess free volume and $1/M$ contributions. The resulting theory is termed the basic lattice cluster theory (BLCT). We now describe the main physical content of the BLCT and its relation to previous theories of copolymer systems.

8.1 The BLCT: General Assumptions

The BLCT is a molecular generalization of random copolymer Flory–Huggins (FH) theory [88] and is based on two significant improvements. The first improvement lies in the description of the polymer–polymer interactions in terms of the united atom groups that are used to represent the monomer structures, an approach that is chemically more realistic than the traditional FH model that is based on monomer–monomer interactions between united monomer groups. The second modification consists in the addition of *entropic* contributions [52] to the interaction parameters χ_{ij}. These temperature-independent portions of χ_{ij} are completely determined within the infinite pressure, fully flexible, long chain limit by the monomer molecular structures and the copolymer compositions [95]. The *entropic* contributions to the χ_{ij} are evaluated simply by counting the numbers of various types of united atom groups (e.g., tri- and tetrafunctional groups) in the monomer structures of the two copolymer blend components and by averaging these numbers over the monomer distributions in the copolymers (see below) [52, 95].

8.1.1
The BLCT Free Energy Expression

The specific Helmholtz free energy of mixing f for a binary incompressible A_xB_{1-x}/C_yD_{1-y} copolymer blend emerges from the BLCT in the apparently familiar form [95]

$$\frac{f}{kT} = \frac{\varphi_1}{M_1}\ln\varphi_1 + \frac{\varphi_2}{M_2}\ln\varphi_2 + \varphi_1^2\chi_{11} + \varphi_2^2\chi_{22} + \varphi_1\varphi_2\chi_{12}\,, \tag{22}$$

where M_i is the number of united atom groups in a single chain of blend component i and φ_i is its volume fraction. The assumption of blend incompressibility implies that $\varphi_1 + \varphi_2 = 1$. The chain occupancy indices M_1 and M_2 are simply related to the average numbers n_α of α-type monomers in individual copolymer chains by

$$M_1 = n_A s_A + n_B s_B\,, \quad M_2 = n_C s_C + n_D s_D\,. \tag{22a}$$

The interaction parameters χ_{ij} of Eq. 22 are given as

$$\chi_{11} = -\frac{z}{2}\left[\frac{\varepsilon_{AA}}{kT}m_A^2 + \frac{\varepsilon_{BB}}{kT}m_B^2 + 2\frac{\varepsilon_{AB}}{kT}m_A m_B\right] - \frac{r_1^2}{z^2}\,, \tag{22b}$$

$$\chi_{22} = -\frac{z}{2}\left[\frac{\varepsilon_{CC}}{kT}m_C^2 + \frac{\varepsilon_{DD}}{kT}m_D^2 + 2\frac{\varepsilon_{CD}}{kT}m_C m_D\right] - \frac{r_2^2}{z^2}\,, \tag{22c}$$

and

$$\chi_{12} = -\frac{z}{2}\left[\frac{\varepsilon_{AC}}{kT}m_A m_C + \frac{\varepsilon_{AD}}{kT}m_A m_D + \frac{\varepsilon_{BC}}{kT}m_B m_C + \frac{\varepsilon_{BD}}{kT}m_B m_D\right] - \frac{2r_1 r_2}{z^2}\,, \tag{22d}$$

where the $\varepsilon_{\alpha\beta}(\alpha, \beta \equiv A, B, C, D)$ are monomer averaged nearest-neighbor van der Waals attractive energies between united atom groups of species α and β and z is the lattice coordination number (taken as six for a cubic lattice). The ratios $m_A = n_A s_A/M_1$, $m_B = n_B s_B/M_1$, $m_C = n_C s_C/M_2$, and $m_D = n_D s_D/M_2$ can conveniently be expressed in terms of the monomer site occupancy numbers s_α and the copolymer compositions $x = n_A/(n_A + n_B)$ and $y = n_C/(n_C + n_D)$ as

$$m_A = 1 - m_B = \frac{x s_A}{x s_A + (1-x)s_B} \tag{23a}$$

and

$$m_C = 1 - m_D = \frac{y s_C}{y s_C + (1-y)s_D}\,. \tag{23b}$$

The entropic structural parameters r_1 and r_2 in Eqs. 22b–22d are the temperature independent portions of χ_{ij} that arise from chains of blend components 1 and 2, respectively. The r_1 and r_2 are weighted averages [95] of the corresponding parameters r_α for the individual pure species α

$$r_1 = r_A \frac{x s_A}{s_1} + r_B \frac{(1-x)s_B}{s_1} \tag{23c}$$

and

$$r_2 = r_C \frac{y s_C}{s_2} + r_D \frac{(1-y)s_D}{s_2}, \tag{23d}$$

where

$$s_1 = x s_A + (1-x)s_B \tag{23e}$$

and

$$s_2 = y s_C + (1-y)s_D \tag{23f}$$

denote the average monomer site occupancy indices for the A_xB_{1-x} and C_yD_{1-y} copolymer chains, respectively. The quantities r_α ($\alpha \equiv$ A, B, C, D) are computed by counting the numbers of tri-$\left(s_\alpha^{(\text{tri})}\right)$ and tetrafunctional $\left(s_\alpha^{(\text{tetra})}\right)$ units in single monomers of species α:

$$r_\alpha = \frac{s_\alpha + s_\alpha^{(\text{tri})} + 3s_\alpha^{(\text{tetra})}}{s_\alpha}. \tag{23g}$$

The scaling factor of $z/2$ in Eqs. 22b–22d is the same as in standard FH theory and formally should be replaced within the BLCT by $(z-2)/2$. Introducing this correction is unnecessary in the BLCT since it merely alters the values of the $\varepsilon_{\alpha\beta}$ that are generally treated as adjustable parameters. The correction, however, becomes relevant when energetic contributions to χ_{ij} are retained beyond the leading energy terms that are included in Eqs. 22b–22d.

8.1.2
The SANS χ_{SANS} Parameter in the BLCT

The small angle neutron scattering (SANS) monomer–monomer interaction parameter χ_{SANS} is defined in terms of the derivative of the specific free energy f as

$$\chi_{\text{SANS}} = C\frac{1}{2}\left[\frac{1}{M_1\varphi_1} + \frac{1}{M_2\varphi_2} - \left.\frac{\partial^2(f/kT)}{\partial\varphi_1^2}\right|_{V,T}\right]. \tag{24a}$$

The factor of C equals $(s_1s_2)^{1/2}$ or $s_1s_2/[s_1(1-\varphi)+s_2\varphi]$ (as explained in Eq. 9) converts the SLCT united atom group–united atom group interaction parameter into the monomer–monomer interaction parameter χ_{SANS}. After some lengthy algebra, Eq. 20a can be transformed into [95]

$$\chi_{SANS} = C\left\{\frac{|r_1 - r_2|^2}{z^2}\right\} + \frac{C}{s_1 s_2}\Big[\chi_{AC} x y s_A s_C + \chi_{BC}(1-x) y s_B s_C \\ + \chi_{AD} x(1-y) s_A s_D + \chi_{BD}(1-x)(1-y) s_B s_D \\ - \chi_{AB} x(1-x) s_A s_B s_2/s_1 - \chi_{CD} y(1-y) s_C s_D s_1/s_2\Big] , \tag{24b}$$

where the binary interaction parameters are defined as

$$\chi_{\alpha\beta} = \frac{z}{2kT}[\varepsilon_{\alpha\alpha} + \varepsilon_{\beta\beta} - 2\varepsilon_{\alpha\beta}] , \quad \alpha \neq \beta \in \{A, B, C, D\} . \tag{24c}$$

If the temperature-independent contribution to χ_{SANS} in Eq. 24b is ignored and each monomer is treated as an entity occupying a single lattice site (i.e., $s_A = s_B = s_C = s_D = 1$), Eq. 24b reduces to the well-known random copolymer FH expression [88]

$$\chi_{FH} = \chi_{AC} xy + \chi_{BC}(1-x)y + \chi_{AD} x(1-y) + \chi_{BD}(1-x)(1-y) \\ - \chi_{AB} x(1-x) - \chi_{CD} y(1-y) . \tag{25}$$

An additional dependence on the copolymer compositions x and y beyond that predicted by simple random copolymer FH theory appears in Eq. 24b through the front conversion factor $C/s_1 s_2$ as well as through factors of s_1/s_2 and s_2/s_1. The main novel feature of Eq. 24b, however, lies in the presence of the temperature-independent portion of χ_{SANS}. This term represents the influence of monomer structural asymmetry on the nonrandom chain packing and coincides with the leading contribution of the LCT for binary blends in the incompressible, high molecular weight, athermal, fully flexible chain limit.

As illustrated in the next subsections, the BLCT has successfully been applied to a wide variety of copolymer blends whose thermodynamics cannot be described by random copolymer FH theory. While both theories share the inability to distinguish between random, diblock, or alternating copolymers of the same composition, the sensitivity of computed thermodynamic properties to the detailed monomer molecular structure of the blend components makes the BLCT a much more powerful theoretical tool than random copolymer FH theory. The relative simplicity of Eq. 24b and the readily computed entropic factors r_1 and r_2 render the BLCT useful for interpreting thermodynamic data and for generating new predictions.

8.1.3 Special Limits of A_xB_{1-x}/C_yD_{1-y} Copolymer Blends

Equation 24b can be used to evaluate χ_{SANS} for special cases of A_xB_{1-x}/C_yD_{1-y} systems. For instance, the interaction parameter χ_{SANS} for a binary blend A_xB_{1-x}/A_yB_{1-y} of two copolymers with the same monomer species but different compositions emerges directly from Eq. 24b by replacing the subscripts C by A and D by B:

$$\chi_{\mathrm{SANS}} = C\left\{\frac{|r_1 - r_2|^2}{z^2}\right\} + \frac{C}{s_1 s_2}\chi_{\mathrm{AB}}(x-y)^2(s_{\mathrm{A}}s_{\mathrm{B}})^2 , \tag{26a}$$

where

$$s_1 = xs_{\mathrm{A}} + (1-x)s_{\mathrm{B}} , \tag{26b}$$

$$s_2 = ys_{\mathrm{A}} + (1-y)s_{\mathrm{B}} , \tag{26c}$$

$$r_1 = r_{\mathrm{A}}\frac{xs_{\mathrm{A}}}{s_1} + r_{\mathrm{B}}\frac{(1-x)s_{\mathrm{B}}}{s_1} , \tag{26d}$$

and

$$r_2 = r_{\mathrm{A}}\frac{ys_{\mathrm{A}}}{s_2} + r_{\mathrm{B}}\frac{(1-y)s_{\mathrm{B}}}{s_2} , \tag{26e}$$

with $r_\alpha(\alpha = \mathrm{A}, \mathrm{B})$ denoting the pure species α entropic parameters given by Eq. 23g. When the monomer site occupancy indices are set to unity ($s_{\mathrm{A}} = s_{\mathrm{B}} = 1$), the temperature-dependent portion of χ_{SANS} from Eq. 26a reduces to the well-known FH random copolymer theory form [88]

$$\chi_{\mathrm{FH}} = \chi_{\mathrm{AB}}(x-y)^2 . \tag{27}$$

The χ_{SANS} parameter for a mixture of a pure homopolymer, say *A*, with a copolymer $\mathrm{C}_y\mathrm{D}_{1-y}$ likewise follows from Eq. 24b by introducing the substitutions $\mathrm{B} \equiv \mathrm{A}$ and $x = 1$. Thus, we find

$$\chi_{\mathrm{SANS}} = C\left\{\frac{|r_1 - r_2|^2}{z^2} + \frac{1}{s_2}\left[\chi_{\mathrm{AC}}ys_{\mathrm{C}} + \chi_{\mathrm{AD}}(1-y)s_{\mathrm{D}} - \chi_{\mathrm{CD}}y(1-y)s_{\mathrm{C}}s_{\mathrm{D}}/s_2\right]\right\} \tag{28a}$$

with $s_1 = s_{\mathrm{A}}$,

$$r_1 = \left[s_{\mathrm{A}} + s_{\mathrm{A}}^{(\mathrm{tri})} + s_{\mathrm{A}}^{(\mathrm{tetra})}\right]/s_{\mathrm{A}} , \tag{28b}$$

and r_2 given by Eq. 23d. The energetic portion of χ_{SANS} in Eq. 28a reduces to the well-known FH random copolymer theory expression [89]

$$\chi_{\mathrm{FH}} = \chi_{\mathrm{AC}}y + \chi_{\mathrm{AD}}(1-y) - \chi_{\mathrm{CD}}y(1-y) \tag{28c}$$

for an $\mathrm{A}/\mathrm{C}_y\mathrm{D}_{1-y}$ system when monomers of species A, C, and D all are assumed to occupy identical volumes.

Equation 24b can similarly be used to generate the χ_{SANS} parameter for a binary homopolymer blend, say A/C, by replacing subscript B with A and D with C. The resulting expression,

$$\chi_{\mathrm{SANS}} = C\left\{\frac{|r_{\mathrm{A}} - r_{\mathrm{C}}|^2}{z^2} + \frac{z}{2kT}(\varepsilon_{\mathrm{AA}} + \varepsilon_{\mathrm{CC}} - 2\varepsilon_{\mathrm{AC}})\right\} , \tag{28d}$$

is a simplified form of the SLCT Eq. 10. A comparison of Eqs. 28d and 10 indicates that the SLCT reduces to the BLCT when the monomer-structure-dependent correction to the interaction energy term is neglected in the SLCT Eq. 9a. As mentioned earlier, the factor of $z/2$ in Eq. 28d should formally be replaced by $(z-2)/2$, but this difference is irrelevant when the interaction energies are treated as adjustable parameters.

8.2 Application of the BLCT to Norbornene/Ethylene Copolymer Blends

The observation by Delfolie et al. [96] that the miscibility of norbornene-co-ethylene (N_xE_{1-x}/N_yE_{1-y}) binary copolymer blends improves significantly when both components are rich in norbornene, i.e., when $x > 1/2$ and $y > 1/2$, stands in sharp contrast to the predictions (Eq. 27) of random copolymer FH theory [88], which implies that the blends are miscible when the composition difference $|x - y|$ is less than a critical constant value. The general theory of Sect. 8.1 that has been used to explain this observation is described elsewhere in some detail [96]. Here we summarize the main results of this theoretical analysis to illustrate some computational aspects of the BLCT.

Due to restrictions imposed by the geometry of a simple cubic lattice and due to the inability [18, 20] of the current LCT to treat polymer systems whose monomers contain closed rings, the cyclic olefin norbornene monomer is modeled by an approximate compact structure (Fig. 10) that is composed of seven united atom groups ($s_A = 7$). The ethylene monomer is represented by a two-bead dimer with two CH_2 groups, i.e., $s_B = 2$. Thus, this simple model incorporates the actual size disparity between these two types of monomers, a feature that influences the magnitude of the *entropic* component χ_s of the interaction parameter χ_{SANS}, and it also affects the dependence of the energetic portion of χ_{SANS} on the compositions x and y of the two blend components.

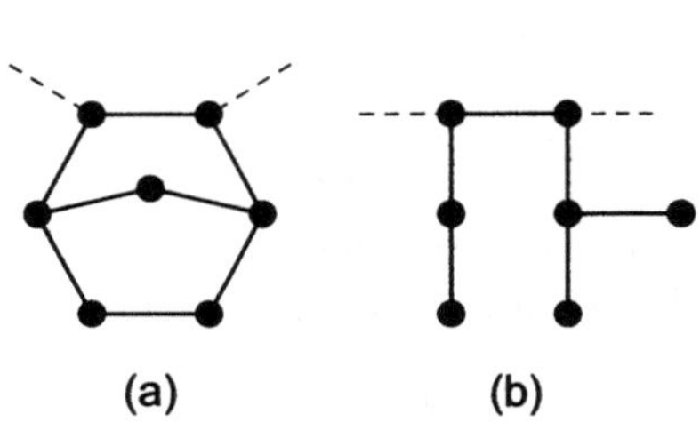

Fig. 10 United atom group model (**a**) of the actual norbornene monomer and (**b**) the simplified norbornene structure used in the BLCT calculations. The *dashed lines* represent the bonds linking the monomer to its neighbors along the chain

The average monomer site occupancy indices s_1 and s_2 of Eqs. 26b and 26c for the two blend components are

$$s_1 = 7x + 2(1-x) \tag{29a}$$

and

$$s_2 = 7y + 2(1-y)\,. \tag{29b}$$

Because each norbornene monomer in our model has three trifunctional united atom CH_n groups and because both norbornene and ethylene units have no tetrafunctional carbons, $r_A = (7+3)/7$ and $r_B = 2/2$. Hence, the partial entropic structure factors r_1 and r_2 of Eqs. 26d and 26e take the form

$$r_1 = \frac{(7+3)x + 2(1-x)}{7x + 2(1-x)} \tag{30a}$$

and

$$r_2 = \frac{(7+3)y + 2(1-y)}{7y + 2(1-y)}\,. \tag{30b}$$

Since SANS experiments have not been performed for the norbornene-co-ethylene mixtures, the conversion factor C for transforming the BLCT χ_{SANS} parameter to a monomer–monomer interaction is taken as $(s_1 s_2)^{1/2}$, which is the more commonly used form. Thus, the entropic portion χ_s of χ_{SANS} becomes

$$\chi_s = (s_1 s_2)^{1/2} \left\{ \frac{|r_1 - r_2|^2}{z^2} \right\} \tag{31a}$$

$$= [(2+5x)(2+5y)]^{1/2} \frac{1}{z^2} \left[\frac{6(x-y)}{(2+5x)(2+5y)} \right]^2 . \tag{31b}$$

According to Eq. 26a, the temperature-dependent portion of χ_{SANS} is given by

$$\chi_h = (s_1 s_2)^{1/2} \chi_{AB} (x-y)^2 (s_A s_B)^2 , \tag{32a}$$

with

$$\begin{aligned}\chi_{AB} &= \frac{z}{2kT}[\varepsilon_{AA} + \varepsilon_{BB} - 2\varepsilon_{AB}] \\ &= \frac{z}{2kT}[\varepsilon_{N\text{-}N} + \varepsilon_{E\text{-}E} - 2\varepsilon_{E\text{-}N}] \equiv \frac{z\varepsilon}{2kT} ,\end{aligned} \tag{32b}$$

where ε is the binary E/N homopolymer blend exchange interaction energy. Combining Eqs. 29a, 29b, 32a, and 32b produces the final expression for χ_h as

$$\chi_h = [(2+5x)(2+5y)]^{1/2} \left[\frac{14(x-y)}{(2+5x)(2+5y)} \right]^2 \frac{z\varepsilon}{2kT} . \tag{32c}$$

The total interaction parameter χ_{SANS} is the sum of the entropic and energetic contributions from Eqs. 31b and 32c. Figure 11 depicts the computed χ_{SANS} parameter as a function of the composition x of component 1 ($N_x E_{1-x}$).

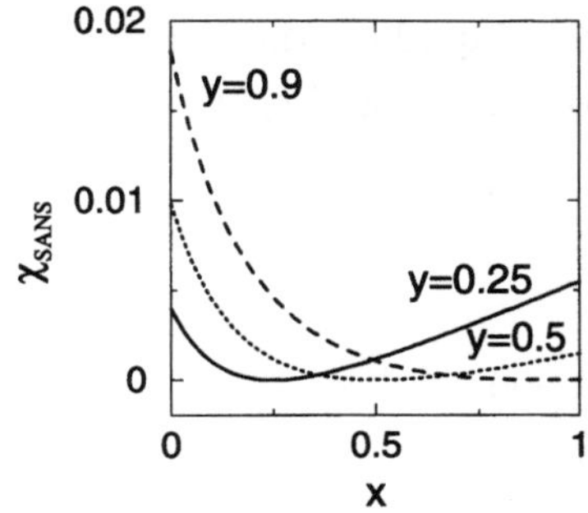

Fig. 11 Interaction parameter χ_{SANS} for ethylene-norbornene copolymer blends computed from Eqs. 31b and 32c as a function of the composition of component 1 (N_xE_{1-x}). Different curves correspond to various compositions y of component 2 (N_yE_{1-y}). The dimensionless exchange energy ε is chosen as $\varepsilon/kT = 10^{-4}$, and $z = 6$. Note that the computed χ_{SANS} is independent of blend composition within the BLCT

The curves correspond to different compositions y of component 2 (N_yE_{1-y}). When both x and y are large, χ_{SANS} indeed becomes small, suggesting a better miscibility of N_xE_{1-x} and N_yE_{1-y} random copolymers, in accord with the experimental observation of Delfolie et al. [96].

8.2.1 Influence of Stiffness and Steric Interactions

Equations 31b and 32c have been derived assuming that all bonds in the random copolymer chains are completely flexible. However, the actual norbornene monomers are rather rigid objects whose steric interactions are considerable. As mentioned in Sect. 6, both internal monomer rigidity and steric interactions influence the partial structural parameters r_i. The rigidity of norbornene monomers (Fig. 10) is introduced into our model by taking the two pairs of side group bonds as completely rigid. The parameter r_i for an A_xB_{1-x} random copolymer chain whose monomers of type A and B have $n_{2,\mathrm{stiff}}^{(A)}$ and $n_{2,\mathrm{stiff}}^{(B)}$ stiff pairs of bonds, respectively, is given by the general formula [96]

$$r_i = \frac{\left[s_A + s_A^{(\mathrm{tri})} + 3s_A^{(\mathrm{tetra})} - n_{2,\mathrm{stiff}}^{(A)}\right]x}{s_A x + s_B(1-x)} + \frac{\left[s_B + s_B^{(\mathrm{tri})} + 3s_B^{(\mathrm{tetra})} - n_{2,\mathrm{stiff}}^{(B)}\right](1-x)}{s_A x + s_B(1-x)}, \quad i = 1, 2, \tag{33a}$$

which is a direct extension of Eq. 23g to copolymer chains composed of monomers with some rigid bonds. Notice that the presence of rigid bonds in the side groups of an α-type monomer always causes a diminution of r_α for that monomer. Specializing Eq. 33a to the N_xE_{1-x}/N_yE_{1-y} species (i.e., setting $n^{(A)}_{2,\mathrm{stiff}} = 2$ and $n^{(B)}_{2,\mathrm{stiff}} = 0$) yields r_1 and r_2 as

$$r_1 = \frac{(7+3-2)x + 2(1-x)}{7x + 2(1-x)} \tag{33b}$$

and

$$r_2 = \frac{(7+3-2)x + 2(1-y)}{7y + 2(1-y)}\,, \tag{33c}$$

which are smaller than the corresponding r_i parameters of Eqs. 30a and 30b for fully flexible chains. Consequently, the corresponding entropic portion χ_s of χ_{SANS} decreases compared to χ_s of Eq. 31b, thereby leading to improved miscibility of norbornene-ethylene copolymer blends. (Introducing rigidity of the norbornene monomers does not affect the calculated temperature dependent portion χ_h of χ_{SANS} in the BLCT, so that Eq. 32c remains valid.)

The bulkiness of the actual norbornene ring implies the presence of the strong steric interactions between bonded norbornene units. These steric interactions have been modeled [96] by treating the bonds connecting adjacent norbornene monomers as semiflexible. The structural parameter $r_i^{(\mathrm{semi})}$ for A_xB_{1-x} random copolymer chains with n_{AA} semiflexible bonds between successive monomers of species A emerges from the BLCT as [96]

$$\begin{aligned} r_i^{(\mathrm{semi})} = & \frac{\left[s_A + s_A^{(\mathrm{tri})} + 3s_A^{(\mathrm{tetra})} - n^{(A)}_{2,\mathrm{stiff}}\right] x}{s_A x + s_B(1-x)} \\ & + \frac{\left[s_B + s_B^{(\mathrm{tri})} + 3s_B^{(\mathrm{tetra})} - 2n^{(B)}_{2,\mathrm{stiff}}\right](1-x)}{s_A x + s_B(1-x)} \\ & + \frac{2(g_A - 1)n_{AA}/(n_A + n_B)}{s_A x + s_B(1-x)}\,, \end{aligned} \tag{34}$$

where

$$g_A = \frac{z}{z - 1 + \exp\left(E_b^{(A)}/kT\right)} \tag{34a}$$

and $E_b^{(A)}$ is the trans–gauche conformational energy for a pair of bonds containing a bond between successive norbornene monomers. Equation 34 becomes identical to Eq. 33a as it must when $g_A = 1$ (i.e., when all backbone bonds are assumed to be fully flexible). Note that the quantity n_{AA} is sequence dependent: n_{AA} vanishes for a perfectly alternating $(AB)_n$ copolymer, equals $(n_A - 1)$ for A_nB_m diblock copolymer, and reduces to $n_A(n_A - 1)/(n_A + n_B)$ for a statistically random copolymer composed of n_A and n_B monomers of types A and B, respectively. The presence of the stiffness factor g_A in Eq. 34 implies that χ_s now depends on temperature, so the term *entropic* χ is no longer appropriate. Because $g_A < 1$ by definition, an increase in the stiffness of the bonds connecting norbornene monomers is predicted to improve the miscibility of norbornene-co-ethylene blends. A more detailed theoretical analysis [96] indicates that the entropic χ_s exerts a dominant role in influencing the miscibilities of the norbornene-co-ethylene blends. A very good representation of the experimental data is obtained when either the rigidity of the norbornene side group or the semiflexibility of the bond between adjacent norbornene monomers is described within the BLCT [96].

8.3 Isotopic Mixtures of Saturated Polybutadienes (sPB)

The simplest A_xB_{1-x}/C_yD_{1-y} blends, perhaps, are those in which the A_xB_{1-x} and C_yD_{1-y} components differ only in the percentage of hydrogen and deuterium atoms [and, of course, in copolymer composition and blend compositions (φ_1, φ_2)]. An example of these systems is provided by mixtures of hydrogenated and deuterated saturated poly(butadiene) (sPB) polymers. Both blend components are random copolymers because of the presence of varying degrees of 1,2 and 1,4 monomers. These weakly interacting isotopic sPB systems have been extensively investigated by SANS techniques [28] and by a nuclear reaction analysis [97]. The same blend samples are used in both experiments and cover a wide range of copolymer compositions (x, y).

Scheffold et al. [97] claim general agreement in the overall values of χ between these two very different measurements, but some systematic discrepancies exist. In particular, fits of the SANS and nuclear reaction analysis data to the form $\chi_{SANS} = \chi_s + \chi_h'/T$ yield rather different χ_s and χ_h' for the three binary blends probed by both methods. (See Table 1 for three examples.) While the temperature-independent portion χ_s of χ_{SANS}, as determined by Scheffold el al. [97], is generally rather small (on the order of 10^{-5}) with both positive and negative signs, the SANS values of χ_s are always negative and large ($-10^{-3} < \chi_s < -10^{-4}$). Also, significant differences (up to 250%) appear in the temperature dependence of the χ parameters (i.e., in the coefficients χ_h') obtained by these two experimental methods. The disparities in χ_s and in χ_h' between the data sets (Table 1) represent one source of uncertainty. Hence, we should not attempt to obtain better agreement between theory and experiment than the magnitude of these discrepancies and/or the stated experimental errors. Another possible source of uncertainty arises from the method of preparation of the samples. In order to have matched pairs of hydrogenated and deuterated sPB chains for the determination of radii of gyration, etc., the same unsaturated perhydro PB samples are then either hydrogenated or deuterated. Unfortunately, the deuteration process is not simple, and a certain degree of H/D scrambling occurs. Thus, the deuterium content varies in unknown fashion among the individual monomers of a single chain and also within the united atom CH_n groups of each monomer.

Table 1 Examples of disparate values of the interaction parameter $\chi_{SANS} = \chi_s + \chi'_h/T$ for saturated PB blends as obtained from SANS [28] and nuclear reaction analysis [97] experiments for identical blend samples

	H52/D66		H66/D52		H88/D78	
	$\chi_s \times 10^4$	χ'_h[K]	$\chi_s \times 10^4$	χ'_h[K]	$\chi_s \times 10^4$	χ'_h[K]
Graessley et al.	– 13.6	1.12	– 6.57	0.656	– 8.17	0.732
Scheffold et al.	4.56	0.428	– 1.22	0.458	-9.28×10^{-3}	0.469

Since the SANS scattering intensity is generally described as depending, in part, on differences in the interaction energies $\{\varepsilon_{\alpha\beta}\}$ of the hydrogenated and deuterated species, the random partial deuterium substitution leads to an additional interaction randomness that cannot presently be treated with the theory developed here. (The random partial deuteration presumably likewise affects the nuclear reaction analysis data.)

The BLCT is applied to isotopic sPB blends by fitting the adjustable parameters of the theory (Sect. 8.3.2) to the two separate data sets generated from the two experimental techniques. First, we describe the fits to the SANS data of Graessley et al. [28]. For notational specificity, let component 1 designate an A_xB_{1-x} random copolymer with a fraction x of hydrogenated 1,2 units and a fraction $(1 - x)$ of hydrogenated 1,4 units ($A \equiv$ H1,2 and $B \equiv$ H1,4), whereas component 2 denotes a statistical copolymer C_yD_{1-y} of partially deuterated 1,2 and 1,4 PB monomers ($C \equiv$ D1,2 and $D \equiv$ D1,4). The united atom monomer structures for both 1,2 and 1,4 monomers are depicted in Fig. 12. The eight adjustable interaction energies that determine χ'_h for A_xB_{1-x}/C_yD_{1-y} binary systems (Eqs. 24b and 24c) can be reduced to a few for the isotopic sPB mixtures by introducing simple assumptions described below. The monomer structures for 1,2 and 1,4 sPB in Fig. 12 indicate that $s_A = s_B = s_C = s_D = 4$, $s_A^{(tri)} = s_C^{(tri)} = 1$, $s_B^{(tri)} = s_D^{(tri)} = 0$, and $s_A^{(tetra)} = s_C^{(tetra)} = s_B^{(tetra)} = s_D^{(tetra)} = 0$, and, consequently, the average monomer site indices s_i

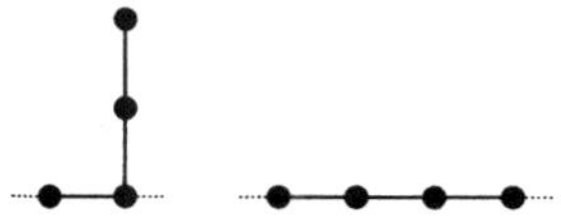

Fig. 12 United atom group models for 1,2 and 1,4 poly(butadiene) (PB) monomers. The models do not distinguish between saturated and unsaturated PB or between cis and trans 1,4 units

$(i = 1, 2)$ in the copolymer occupancy indices coincide with those corresponding to the individual monomers of either species:

$$s_1 = 4x + 4(1 - x) = 4$$

and

$$s_2 = 4y + 4(1 - y) = 4\,.$$

The partial entropic structural parameters r_1 and r_2 of Eqs. 23c and 23d now simplify to

$$r_1 = \frac{(4+1)x + 4(1-x)}{4x + 4(1-x)} = \frac{x+4}{4} \tag{35a}$$

and

$$r_2 = \frac{(4+1)y + 4(1-y)}{4y + 4(1-y)} = \frac{y+4}{4}\,. \tag{35b}$$

In order to reduce the number of adjustable parameters to the bare minimum, all heterocontact interaction energies $\{\varepsilon_{\alpha\beta}\}(\alpha \neq \beta)$ are assumed to satisfy the Berthelot combining rule [98], i.e.,

$$\varepsilon_{\alpha\beta} = (\varepsilon_{\alpha\alpha}\varepsilon_{\beta\beta})^{1/2}\,, \quad \alpha, \beta = \mathrm{A, B, C, D}\,, \quad \alpha \neq \beta\,. \tag{36a}$$

The second simplifying assumption involves the use of the polarizability model of Bates et al. [99], which postulates a simple scaling relation between the interaction energies for two hydrogenated and two deuterated monomer groups:

$$\varepsilon_{\mathrm{CC}} = \gamma^2 \varepsilon_{\mathrm{AA}} \tag{36b}$$

and

$$\varepsilon_{\mathrm{DD}} = \gamma^2 \varepsilon_{\mathrm{BB}}\,, \tag{36c}$$

where the scaling factor γ is less than unity to reflect the weaker attraction between two partially deuterated species. Again, to minimize the number of parameters, a common value of γ is assigned for both 1,2 and 1,4 monomers. This assumption also ignores the fact that the experimental samples have varying degrees of partial deuteration and should therefore be analyzed within the BLCT by allowing the γ factors (and scattering lengths) to vary with the monomers' deuterium content. As already mentioned, a theoretical treatment of variable deuteration poses serious technical difficulties.

8.3.1
The χ_{SANS} Parameter

The small angle neutron scattering χ_{SANS} parameter of Eq. 24b can be converted by using Eqs. 35a–36c to the rather compact expression

$$\chi_{SANS} = \frac{(x-y)^2}{4z^2} + \frac{2z}{kT}\left\{(1-\gamma)\left[y\varepsilon_{AA}^{1/2} + (1-y)\varepsilon_{BB}^{1/2}\right] + (x-y)\left[\left(\varepsilon_{AA}^{1/2} - \varepsilon_{BB}^{1/2}\right)\right]\right\}^2 , \tag{37}$$

which contains only three adjustable parameters: two self-interaction van der Waals energies ε_{AA} and ε_{BB} for the 1,2 and 1,4 perhydrogenated species, respectively, and the polarizability scaling factor γ. The first term in braces in Eq. 37 dominates over the second, but the hydrogen/deuterium swap effect is lost without this second term, a feature consistent with the nature of the swap phenomenon that arises from *very small* differences between the two isotopic species. Note that the energetic portion of the χ_{SANS} from Eq. 37 exhibits a structure similar to the form

$$\chi_\delta = \frac{v_0}{2kT}(\delta_1 - \delta_2)^2 \tag{38}$$

that establishes the basis of solubility parameter solution theory and that has been used by Graessley, Lohse, and coworkers [27–32] to describe their experimental data for χ_{SANS} in terms of the pure component solubility parameters δ_1 and δ_2. Because the squares of the solubility parameters are *by definition* proportional to the cohesive energy densities of the pure melts, the solubility parameters rigorously do not contain contributions from the athermal packing entropy that are embodied in the entropic portion of χ in the BLCT. Thus, the use of Eq. 38 to interpret experimental χ_{SANS} data represents theoretically unjustified empiricism when the χ_{SANS} parameter has a significant temperature-independent component of pure entropic origin. The lack of a separate entropic portion χ_s in solubility parameter theory is a serious deficiency of that approach. An improved version of solubility parameter theory could be formulated, for instance, by merging Eq. 38 with the SLCT Eq. 11b.

8.3.2
Comparison with Experimental Data

The three parameters in Eq. 37 are fit to experimental $\chi_{SANS}(T)$ data for a series of ten isotopic sPB random copolymer blends that differ in the compositions x and y of the two blend components. The values of the fitting parameters are, however, not unique, and many sets of ε_{AA}, ε_{BB}, and γ reproduce the data with similar accuracy and standard deviations. Among the many sets, we choose the following: $\varepsilon_{AA} \equiv \varepsilon_{1,2-1,2} = 238.3$ K, $\varepsilon_{BB} \equiv \varepsilon_{1,4-1,4} = 240.8$ K, and $\gamma = 0.9901$. The value of γ agrees with the polarizability model prediction [99] of $\alpha_{C\text{-}D}/\alpha_{C\text{-}H} = 0.98\text{–}0.99$. Moreover, the higher self-interaction energy $\varepsilon_{\alpha\alpha}$ for the less branched 1,4 polybutadiene species ($\varepsilon_{1,4\text{-}1,4} > \varepsilon_{1,2\text{-}1,2}$) is consistent with our previous fits [80] of the LCT to equation of state (i.e., PVT) data for polyolefin melts as well as with the analysis of Schweizer and coworkers [100–102].

Figure 13 illustrates the comparison between the experimental $\chi_{SANS}(T)$ data for a pair of systems that are labeled by Graessley et al. [28] as H38/D25 and H25/D38 and the BLCT $\chi_{SANS}(T)$ as calculated from Eq. 37 for the set of ε_{AA}, ε_{BB}, ε_{AB}, and γ specified above. The notation H38/D25, for example, denotes the isotopic blend of hydrogenated (H) and deuterated (D) polybutadiene random copolymers, and the numbers indicate the fractions of 1,2 units in each of the blend components. The experimental points [28] are designated in Fig. 13 by circles and triangles, whereas the lines represent the theoretical fits. Figure 13 demonstrates that the theory reproduces the overall values of the χ parameters within the experimental error bars of $\pm 1.5 \times 10^{-4}$ [28].

Note that in order to describe the $\chi_{SANS}(T)$ data for ten binary sPB blends, we employ only three adjustable parameters, while the analysis of Graessley et al. [28–30] requires a separate set of temperature-dependent solubility parameters for each individual copolymer species (i.e., 20 sets of solubility parameters in all). Despite the huge contraction of the $\delta_1(T)$ and $\delta_2(T)$ values into just the three parameters (ε_{AA}, ε_{BB}, and γ) of the BLCT, the slopes χ'_h of $\chi_{SANS}(T)$ vs. $1/T$, however, differ considerably between theory and experiment. Some decrease in this discrepancy can be obtained by introducing different chain stiffness for the two blend components or by allowing γ to differ slightly for the 1,2 and 1,4 monomer units, but given the experimental

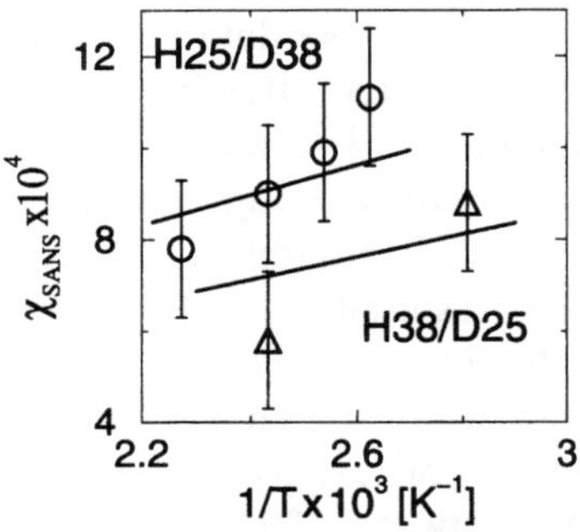

Fig. 13 Interaction parameter χ_{SANS} as a function of inverse temperature $1/T$ for symmetric ($\varphi_1 = \varphi_2 = 0.5$) H25/D38 and H38/D25 isotopic saturated poly(butadiene) blends. *Solid lines* are the BLCT fits to a collective set of experimental data [28] for ten Hx/Dy and Hy/Dx mixtures. *Circles* and *triangles* indicate data for the H25/D38 and H38/D25 mixtures, respectively. The experimental *error bars* for χ_{SANS} are taken as the upper limits for the stated accuracy ($\pm 1.5 \times 10^{-4}$) of the SANS measurements [28]

error bars, there appears to be no point in adding extra parameters to achieve these improvements.

An important feature of the BLCT Eq. 37 is the proper reproduction of the deuterium swap effect, i.e., a larger χ is always obtained when the more branched polyolefin component is deuterated. This deuterium swap effect has been found in numerous experiments [28, 97, 103, 104] and follows as a direct consequence of Eq. 37 provided that the self-interaction energy $\varepsilon_{\alpha\alpha}$ is higher for the less branched polyolefin species. It should be possible to rationalize the latter constraint on the interaction energies from the models employed for off-lattice simulations of alkane systems. These continually refined models use considerably larger Lennard–Jones interaction parameters for CH_3 groups than for CH_n ($n = 0, 1, 2$) units.

One feature contributing to the differences between the experimental and theoretical slopes χ'_h in Fig. 13 is the presence of a rather huge negative temperature-independent portion χ_s of χ_{SANS} for all the binary isotopic sPB blends studied by Graessley et al. [28] (for instance, $\chi_s = -0.0035$ and -0.0022 for the H08/D25 and H25/D08 blends, respectively). In contrast, the computed entropic part χ_s of χ_{SANS} in Eq. 37 is always positive

$$\chi_s = \frac{(x-y)^2}{4z^2}$$

and small. [A slightly negative χ_s can, however, be predicted by the SLCT when chain semiflexibility is included (Sect. 7) or by the LCT that treats the system as compressible.] On the other hand, both the entropic and energetic portions of χ_{SANS} obtained by Graessley et al. [28] for some systems are quite different from those found by Scheffold et al. [97] (Table 1). Therefore, it is also relevant to describe the neutron reaction analysis data with the BLCT.

Fits to the nuclear reaction analysis data [97] for eight binary sPB blends have been performed using the interaction energies $\varepsilon_{AA} = 238.3$ K and $\varepsilon_{BB} = 240.8$ K obtained from fits to the SANS data [28] and by allowing γ to be the only adjustable parameter (although allowing the energies to be adjustable parameters could, in principle, lead to improved fits). As shown in Table 2, theoretical and experimental slopes χ'_h agree well, with the discrepancy exceeding 20% only for the H52/D66 sample. All other samples exhibit quite small values of χ_s that are presumably smaller than the experimental error bars. If we assume that the SANS error bars of $\pm 1.5 \times 10^{-4}$ also represent the accuracy of the nuclear reaction analysis data [97], there is good agreement between the theoretical and experimental [97] χ_{SANS} parameters over a wide temperature range (between 400 and 500 K) for half of the samples considered, but further refinement of the current theoretical model (such as introducing chain semiflexibility or differing γ for the 1,2 and 1,4 monomers) would undoubtedly improve the agreement.

Table 2 SANS interaction parameter $\chi_{SANS} = \chi_s + \chi'_h/T$ of saturated poly(butadiene) blends as determined from the nuclear reaction analysis [97] experiments and from the BLCT fits to the data

Blend	$\chi_s \times 10^4$ Experiment	$\chi_s \times 10^4$ Theory	χ'_h[K] Experiment	χ'_h[K] Theory
H52/D66	3.87	1.36	0.3633	0.518
H66/D52	- 1.20	1.36	0.459	0.412
H66/D75	0.27	0.56	0.410	0.497
H75/D86	0.8	0.84	0.543	0.505
H78/D88	- 0.95	0.69	0.583	0.501
H86/D75	- 0.70	0.84	0.488	0.422
H86/D94	0	0.25	0.647	0.492
H88/D78	- 0.009	0.69	0.469	0.425

8.4 Unsaturated Isotopic Polybutadiene Blends

Blends of protonated (HPB) and perdeuterated (DPB) unsaturated polybutadienes have also been the subject of extensive experimental investigations [53, 87, 106, 107]. Both blend components are typical random copolymers because of the presence of varying degrees of randomly placed 1,2 and 1,4 PB addition units along the polymer chain.

Early studies by Bates et al. [106, 107] and by Sakurai et al. [87] reveal that these systems exhibit an upper critical solution temperature (UCST), i.e., undergo phase separation upon cooling. Subsequent investigations have focused on the combined influence of isotope effects and the blend microstructure on the miscibility patterns of these random copolymer binary mixtures. In particular, a series of systematic SANS experiments by Jinnai et al. [53] demonstrate that the UCST phase behavior that had previously been observed for these systems [87, 106, 107] remarkably converts to a lower critical solution temperature (LCST) phase separation with an increase in the vinyl content of the HPB component when the vinyl content of the DPB component remains fixed. This phenomenon cannot be explained by the traditional extension of FH theory to random copolymers since this theory is derived under the assumption that the individual $\chi_{\alpha\beta}$ are of purely energetic origin. Thus, the FH random copolymer theory [88] is capable, at most, of predicting the conversion of a UCST phase separation system into a completly miscible system.

8.4.1 The χ_{SANS} Parameter

The BLCT is applied [108] to these isotopic PB blends in order to elucidate the molecular factors responsible for the transition from UCST to LCST behavior that occurs when the blend microstructure is changed [53]. The united atom model for the 1,2 and 1,4 PB species does not distinguish between cis and trans 1,4 units, nor between head-to-head and head-to-tail placements of successive 1,4 units, but this deficiency of the model is not considered by us as very serious since these fine details probably exert a secondary influence on the blend compatibility. The BLCT computations are performed assuming the presence of only three independent microscopic van der Waals energies ε_{AA}, ε_{BB}, and ε_{AB} as representing the interactions between two 1,2, two 1,4, and a pair of 1,2 and 1,4 united atom CH_n groups, respectively. The remaining seven energy parameters are determined by using the simple scaling relations in the spirit of the polarizability model of Bates et al. [99]:

$$\varepsilon_{CC} = \gamma^2 \varepsilon_{AA} , \ \varepsilon_{DD} = \gamma^2 \varepsilon_{BB} , \ \varepsilon_{CD} = \gamma^2 \varepsilon_{AB} , \ \varepsilon_{AC} = \gamma \varepsilon_{AA} , \qquad (39)$$
$$\varepsilon_{BD} = \gamma \varepsilon_{BB} , \varepsilon_{BC} = \varepsilon_{AD} = \gamma \varepsilon_{AB} .$$

Given these assumptions, the SANS parameter χ_{SANS} of Eq. 24b takes the form

$$\chi_{SANS} = \frac{(x-y)^2}{4z^2} + \frac{2z}{kT} \left\{ \varepsilon_{AA}(\gamma y - x)^2 + \varepsilon_{BB}[\gamma(1-y) - (1-x)]^2 - 2\varepsilon_{AB}(x - \gamma y)[\gamma(1-y) - (1-x)] \right\} , \qquad (40)$$

where following Jinnai et al. C has been replaced by $s_1 s_2/[s_1(1-\varphi) + s_2\varphi]$ and where x and y are the compositions of 1,2 units in the HPB and DPB components, respectively.

If both blend components are hydrogenated or deuterated polybutadienes (i.e., if $\gamma = 1$), Eq. 40 simplifies to

$$\chi_{SANS} = \frac{(x-y)^2}{4z^2} + \frac{2z}{kT}(x-y)^2[\varepsilon_{AA} + \varepsilon_{BB} - 2\varepsilon_{AB}] . \qquad (41)$$

Equation 41 is a special case of Eq. 26a for a binary A_xB_{1-x}/A_yB_{1-y} mixture of two copolymers that emerges when the monomer of both species A and B have identical site occupancy indices ($s_A = s_B = 4$). Thus, the SANS interaction parameter of the PB A_xB_{1-x}/A_yB_{1-y} blends is proportional to the square of the difference in the copolymer compositions x and y [$\chi_{SANS} \sim (x-y)^2$]. Since, as noted by Schwahn and coworkers [109], the exchange energy $\varepsilon_{AA} + \varepsilon_{BB} - 2\varepsilon_{AB}$ is positive for the PB A_xB_{1-x}/A_yB_{1-y} blends (with any x and y), only UCST phase behavior is possible for these mixtures of two perhydro polybutadiene copolymers. The remarkable conversion from UCST to LCST phase diagrams that has been observed by Jinnai et al. [53] as $|x-y|$ is varied arises within the BLCT from a change of the sign of χ_h with

$|x - y|$. While both the BLCT and random copolymer FH theory [88] admit of a variation of the sign of χ_h with the microstructure of A_xB_{1-x}/C_yD_{1-y} blends, only the BLCT can explain the occurrence of LCST phase behavior because the BLCT provides an estimate of the temperature-independent portion of χ_{SANS} that is absent in FH theory. The example of HPB/DPB isotopic blends demonstrates that the very small energetic difference between the H/D isotopes is sufficient to induce this remarkable change in phase behavior.

8.4.2
Comparison with Experimental Data

The four parameters of Eq. 40 are fit [108] to the experimental data of Jinnai et al. [53] for five isotopic PB random copolymer blend samples with varying compositions x of the protonated PB species and a constant composition y of the deuterated polybutadiene. Table 3 summarizes the theoretical and experimental values of both χ_s and χ_h' and indicates that the theory reproduces semiquantitatively the SANS data [53] in the sense that it precisely describes the variation in the nature of the phase transition (upper vs. lower critical solution temperature phase diagrams) with the microstructure of the isotopic PB blends. The type of phase behavior is correctly predicted [108] by the BLCT for all five PB blends considered. Within the BLCT, an LCST phase diagram for the isotopic PB blends stems from the competition between a negative energetic portion χ_h' and a sufficiently positive entropic part χ_s of χ_{SANS}. This competition is more noticeable by considering the spinodal condition that has the following form in the BLCT:

$$\frac{1}{M_1\varphi} + \frac{1}{M_2(1-\varphi)} - 2\left[\chi_s + \frac{\chi_h'}{T}\right] = 0\,, \tag{42}$$

where φ is the blend composition and M_1 and M_2 are the numbers of united atom groups in the individual chains of species 1 and 2, respectively. When χ_h' is negative, a sufficiently large $\chi_s > 0$ leads to LCST phase behavior.

Except for one blend sample (where both x and y are low and very similar to each other), the experimental and theoretical values of χ_h' accord reasonably well. Good agreement for the temperature-independent portion χ_s of χ_{SANS} is confined, however, to the two samples with the largest compositional difference $\Delta = |y - x|$ between the two blend components and with the largest experimental χ_s. The experimental values of χ_s are negative for the blend samples for which $\Delta = 0.01$ and 0.23, but the flexible chain BLCT predicts χ_s as always positive. This limitation precludes obtaining a perfect fit to the data [53]. However, as discussed above, the more general LCT, which applies to compressible polymer systems with arbitrary molecular weights and chain semiflexibility, can admit a small negative χ_s of a magnitude comparable to those observed for the two samples. When both χ_s and χ_h' are negative, the system must be treated as compressible and possibly the polymers described as semiflexible molecules in order to explain an LCST phase diagram. Another neglected feature that may affect χ_s is the presence of cis-trans randomness of the 1,4 butadiene units, a feature not easily modeled by a simple analytical approach.

The values of the fitting parameters again are not unique. Many sets of ε_{AA}, ε_{BB}, and ε_{AB}, and γ lead to similar results as illustrated by the set in Table 3 that are obtained for $\varepsilon_{AA} \equiv \varepsilon_{1,2-1,2} = 260.1$ K, $\varepsilon_{BB} \equiv \varepsilon_{1,4-1,4} = 271.3$ K, $\varepsilon_{AB} \equiv \varepsilon_{1,2-1,4} = 265.7$ K, and $\gamma = 0.987$. The higher interaction energies for the unsaturated PB blends than the $\{\varepsilon_{\alpha\beta}\}$ determined for saturated PB mixtures (Sect. 8.3.2) are consistent with the presence of double bonds in the unsaturated species. Moreover, the differences between $\varepsilon_{1,2-1,2}$, $\varepsilon_{1,4-1,4}$, and $\varepsilon_{1,2-1,4}$ are indeed small as expected for weakly interacting systems. The value of heterocontact interaction energy $\varepsilon_{1,2-1,4}$ departs from that resulting from the Berthelot geometric combining rule [98] by about 0.02%. Note that forcing $\varepsilon_{1,2-1,4}$ to equal the geometrical mean $(\varepsilon_{1,2-1,2}\varepsilon_{1,4-1,4})^{1/2}$ automatically leads within the BLCT to a positive χ_h' for arbitrary microstructures of the isotopic PB blends and, therefore, always to the prediction of a UCST phase behavior [108].

The review article by Schwahn in this volume describes more recent tests of the BLCT for isotopic PB blends by Schwahn and coworkers [109]. These tests have been applied to data sets roughly three times larger than the data set of Jinnai et al. that is analyzed by us. Moreover, the analysis by Schwahn and coworkers was performed using only three adjustable parameters since one of the self-interaction energies ε_{ii} was determined from fits of the LCT to pure component equation of state (PVT) data. Excellent agreement is found between theory and experimental for both the entropic χ_s and energetic χ_h' portions of the χ_{SANS} parameter.

Table 3 SANS interaction parameter $\chi_{SANS} = \chi_s + \chi_h'/T$ of isotopic DPB(x)/HPB(y) unsaturated poly(butadiene) blends[a]

Blend sample Experiment	$\chi_s \times 10^4$ Theory	$\chi_s \times 10^4$ Experiment	χ_h'[K] Theory	χ_h'[K] Experiment
DPB(16)/HPB(17)	- 4.38	1.36×10^{-2}	0.352	0.784
DPB(16)/HPB(39)	- 5.19	3.80	0.234	0.214
DPB(16)/HPB(52)	3.57	9.30	- 0.144	- 0.128
DPB(16)/HPB(65)	14.6	16.9	- 0.556	- 0.457
DPB(16)/HPB(80)	30.3	29.2	- 0.814	- 0.859

[a] Numbers in parentheses are rounded off values for percentage of 1,2 units in the blend components

9
Order–Disorder Transition in Polystyrene-b-Poly(n-Alkyl Methacrylate) Diblock Copolymers

Since our LCT prediction [110] of the possible existence of a lower order-disorder transition (LDOT) in diblock copolymers, interest in this phenomenon has increased significantly due to experiments of Russell and coworkers [111, 112] that confirm our predictions and due to experiments of Ruzette et al. [113] that reveal the existence of different types of phase behavior in a homologous family of diblock copolymers formed from styrene and a series of *n*-alkyl methacrylates. The extremely well-studied system of PS-b-PMMA [poly(styrene)-b-poly(methyl methacrylate)] is known to undergo an upper order–disorder transition (UODT); the short side chain *n*-alkyl methacrylate copolymers ($n = 2$, 3, and 4) are found to order upon heating (LDOT systems); and diblock copolymers with longer *n*-alkyl group ($n = 6$, 8, 12) are observed to order on cooling. Quite remarkably, the block copolymer, consisting of styrene as one block and the other block a 53–47 weight percent random copolymer of methyl ($n = 1$) and lauryl ($n = 12$) methacrylates [PS-b-P($MMA_{0.74}$-co-$LMA_{0.26}$], orders on heating, despite the fact that both PS-b-PMMA and PS-b-PLMA copolymers individually exhibit classical upper order–disorder transitions! These striking results clearly indicate that the microphase separation in these systems is subtly, but decisively, affected by the monomer molecular structures and the detailed molecular interactions.

9.1
Unified Modeling of the *n*-alkyl Methacrylate Systems

This striking variation in the nature of the order–disorder transition with the length of the alkyl side group in the *n*-alkyl methacrylate (*n*AMA) block component can only be explained by a theory that distinguishes not only between the chemical structures of the styrene and *n*-alkyl methacrylate monomers, but also between different *n*-alkyl methacrylate molecules. Flory–Huggins theory is of no use in these regards since it treats monomers as structureless entities whose only differences are reflected by ascribing different monomer-monomer interaction parameters χ_{ij} to the individual components in the system. The BLCT provides a more realistic molecular description by specifying united atom type structures for individual monomers, by explicitly computing (with no adjustable parameters) the entropic portion of χ from the numbers of tri- and tetrafunctional united atom groups in these monomer structures, and by basing the computation of the temperature-independent portion of χ on the more microscopically faithful interactions between united atom groups. However, in order to minimize the number of adjustable parameters in the BLCT, all united atom groups of a monomer have heretofore been assumed to interact with a common, monomer averaged energy. While the approximation of monomer averaged interaction energies may suffice as a zeroth-order model for some chemically fairly homogeneous systems like polyolefins, the monomer averaged interaction models are clearly inadequate for describing the observed [113] striking variation in the ordering behavior of styrene-b-*n*-alkyl methacrylate diblock copolymers with the length of the alkyl chain. Thus, the BLCT has been extended [114] to treat binary blends in which each pair of chemically distinct united atom groups is permitted to interact with different interaction energies [49].

For simplicity, we invoke the traditional assumption that the SANS interaction parameters for A/B binary blends and A-b-B diblock copolymer melts are identical. The BLCT Eq. 28d for χ_{SANS} of an A/B binary homopolymer blend,

$$\begin{aligned}\chi_{\text{SANS}} &= \chi_{\text{s}} + \chi'_{\text{h}}/T \\ &= C\left\{\frac{|r_1 - r_2|^2}{z^2}\right\} + C[\varepsilon_{\text{AA}} + \varepsilon_{\text{BB}} - 2\varepsilon_{\text{AB}}]\frac{z-2}{kT}\,,\end{aligned} \tag{43}$$

is still valid when a monomer of polymer species A has $s_{\text{A}_1}, s_{\text{A}_2}, \ldots, s_{\text{A}_k}$ united atom groups of type $\text{A}_1, \text{A}_2, \ldots, \text{A}_k$, respectively (where $s_{\text{A}_1} + s_{\text{A}_2} + \ldots + s_{\text{A}_k} = s_{\text{A}}$), and a monomer of polymer species B has $s_{\text{B}_1}, s_{\text{B}_2}, \ldots, s_{\text{B}_l}$ united atom groups of type $\text{B}_1, \text{B}_2, \ldots, \text{B}_k$, respectively ($s_{\text{B}_1} + s_{\text{B}_2} + \ldots + s_{\text{B}_l} = s_{\text{B}}$). However, the effective attractive van der Waals energies ε_{AA}, ε_{BB}, and ε_{AB} are now explicit functions of the specific group site occupancy indices s_{A_i} and s_{B_i} and the specific interaction energies $\varepsilon_{\text{A}_i\text{B}_j}$

$$\varepsilon_{\text{AA}} = \sum_{i=1}^{k} \varepsilon_{\text{A}_i\text{A}_i}\left(\frac{s_{\text{A}_i}}{s_{\text{A}}}\right)^2 + 2\sum_{i=1}^{k}\sum_{j>i}^{k} \varepsilon_{\text{A}_i\text{A}_j}\frac{s_{\text{A}_i}s_{\text{A}_j}}{s_{\text{A}}^2}\,, \tag{44a}$$

$$\varepsilon_{\text{BB}} = \sum_{i=1}^{l} \varepsilon_{\text{B}_i\text{B}_i}\left(\frac{s_{\text{B}_i}}{s_{\text{B}}}\right)^2 + 2\sum_{i=1}^{l}\sum_{j>i}^{l} \varepsilon_{\text{B}_i\text{B}_j}\frac{s_{\text{B}_i}s_{\text{B}_j}}{s_{\text{B}}^2}\,, \tag{44b}$$

$$\varepsilon_{\text{AB}} = \sum_{i=1}^{k}\sum_{j=1}^{l} \varepsilon_{\text{A}_i\text{B}_j}\frac{s_{\text{A}_i}s_{\text{B}_j}}{s_{\text{A}}s_{\text{B}}}\,. \tag{44c}$$

The specific interaction energies $\{\varepsilon_{\alpha_i\beta_j}\}$ in Eqs. 44a–44c are the adjustable parameters of the theory, but, as shown below for the PS-b-PnAMA systems, not all are independent. The factor C in Eq. 43 again ensures that χ is normalized as a traditional monomer–monomer interaction parameter. Equations 43 and 44 can readily be extended to systems in which one component is a copolymer, and the details are given in [114].

The knowledge of both portions of $\chi_{\text{SANS}} = \chi_{\text{s}} + \chi'_{\text{h}}/T$ for the A-b-B diblock copolymer system enables evaluating the order–disorder transition temperature that is generally determined from the vanishing of the inverse structure factor $1/S(q^*)$, where q^* denotes the wavevector for which $S(q)$ has a maximum. A modified version of Leibler theory [115] relates χ and $S(q^*)$ through the expression

$$\frac{1}{S(q^*)} = \frac{F(x_1, x_2, f)}{N} - 2\chi_{\mathrm{SANS}}\,, \tag{45}$$

where N is the polymerization index of the diblock and where the function F is generalized [116] to treat unequal Kuhn lengths for the two segment types,

$$F(x_1, x_2, f) = \frac{f^2 g(x_1) + 2f(1-f)h(x_2) + (1-f)^2 g(x_2)}{f^2(1-f)^2[g(x_1)g(x_2) - h^2(x_1)h^2(x_2)]}\,, \tag{46}$$

with

$$g(x_i) = 2x_i^{-2}[x_i + \exp(-x_i) - 1]\,, \quad i = 1, 2\,,$$
$$h(x_i) = 2x_i^{-1}[1 - \exp(-x_i)]\,, \quad i = 1, 2\,,$$

and

$$x_i \equiv (q^* R_i)^2\,, \quad i = 1, 2\,.$$

In the above equations, the $R_i = (N_i a_i^2/6)^{1/2} (i = 1, 2)$ are the radii of gyration of the blocks, the a_i are the Kuhn lengths, $f \equiv f_1$ is the volume fraction of block 1, and $N_1 + N_2 = N$. The volume fraction f may be determined from the mole fraction $y \equiv y_1 = N_1/N$ and from the monomer molar volumes v_1 and v_2 through the standard relation

$$f = \frac{y v_1}{y v_1 + (1-y) v_2}\,.$$

The order–disorder transition temperature T_{ODT} simply follows from Eqs. 45 and 46 as

$$T_{\mathrm{ODT}} = \frac{\chi_{\mathrm{h}}' N}{(1/2)F(x_1, x_2, f) - \chi_{\mathrm{s}} N}\,. \tag{47}$$

Equation 47 differs from the standard Leibler formula [115] by the presence of the extra term $\chi_{\mathrm{s}} N$ and the more general formulation of the function F as given by Eq. 46.

The expression for the interaction parameter χ_{SANS} in Eq. 43 is specialized below for PS-b-poly(n-alkyl methacrylate) diblock copolymer melts. The monomer structures for both the styrene ($s_{\mathrm{A}} \equiv s_{\mathrm{S}} = 8$) and n-alkyl methacrylate ($s_{\mathrm{B}} \equiv s_{n\text{-AMA}} = 6 + n$) monomers are depicted in Fig. 14. Stiffness in the styrene molecule is introduced into our model by treating the two pairs of styrene side group bonds (Fig. 14) as completely rigid ($n_{2,\mathrm{stiff}}^{(\mathrm{S})} = 2$). In order to minimize the number of adjustable parameters, we distinguish only three types of united atom units in all these diblock copolymer systems considered: alkyl CH_n groups ($n = 0$, 1, 2, or 3), aromatic CH_n groups

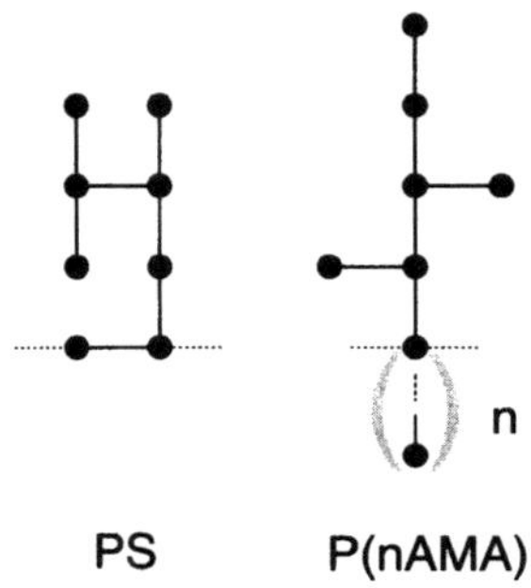

Fig. 14 United atom group structures for the styrene (S) and *n*-alkyl methacrylate (*n*AMA) monomers

($n = 0$ or 1), and oxygen atoms. This assumption formally leads to the presence of six different interaction energies $\varepsilon_{\text{Alk-Alk}}$, $\varepsilon_{\text{Ar-Ar}}$, $\varepsilon_{\text{O-O}}$, $\varepsilon_{\text{Alk-Ar}}$, $\varepsilon_{\text{Alk-O}}$, and $\varepsilon_{\text{Ar-O}}$, but only three combinations of these energies are found [114] to be independent parameters in the BLCT for the systems studied. No interaction is ascribed to the united atom group pairs involving the tetrafunctional carbon of the *n*-alkyl methacrylate monomer due to the obvious steric hindrances.

The entropic and energetic portions of the effective interaction parameter χ_{SANS} for the PS-b-P(*n*-alkyl methacrylate) melt can be expressed finally as [114]

$$\chi_{\text{s}} = [8(6+n)]^{1/2} \frac{1}{z^2} \left(\frac{9}{8} - \frac{10+n}{6+n} \right)^2 \tag{48a}$$

and

$$\chi_{\text{h}}' = [8(6+n)]^{1/2} \left[\frac{z-2}{2k} \frac{1}{16(6+n)^2} \right] \left\{ 9(2+n)^2 \varepsilon_1 + 12(2+n)\varepsilon_2 + 16\varepsilon_3 \right\} , \tag{48b}$$

where the three independent combinations of interaction energies,

$$\varepsilon_1 \equiv \varepsilon_{\text{Alk-Alk}} + \varepsilon_{\text{Ar-Ar}} - 2\varepsilon_{\text{Alk-Ar}} ,$$

$$\varepsilon_2 \equiv 6(\varepsilon_{\text{Ar-Ar}} - \varepsilon_{\text{Alk-Ar}}) + 4(\varepsilon_{\text{Alk-O}} - \varepsilon_{\text{Ar-O}}) ,$$

$$\varepsilon_3 \equiv \varepsilon_{\text{O-O}} + 9\varepsilon_{\text{Alk-Ar}} - 12\varepsilon_{\text{Ar-O}} ,$$

are the only three adjustable parameters of the theory. The corresponding expressions χ_{s} and χ_{h}' for the random copolymer PS-b-P(X_xY_{1-x}) diblock copolymers, with X and Y denoting, respectively, *n*-alkyl methacrylate monomers with $n = p$ and $n' = q$ united atom groups, take the form [114]

$$\chi_s = \{8[6 + q + x(p - q)]\}^{1/2} \frac{1}{z^2} \left[\frac{9}{8} - \frac{10 + q + x(p - q)}{6 + q + x(p - q))} \right]^2 , \quad (49a)$$

$$\chi'_h = \{8[6 + q + x(p - q)]\}^{1/2} \frac{z - 2}{2k} \frac{1}{16[6 + q + x(p - q)]^2} \quad (49b)$$
$$\times \left\{ 9[2 + q + x(p - q)]^2 \varepsilon_1 + 12[2 + q + x(p - q)]\varepsilon_2 + 16\varepsilon_3 \right\} .$$

Setting $x = 0$ (or $x = 1$) in Eqs. 49a and 49b recovers Eqs. 48a and 48b, respectively. When Eqs. 49a and 49b are specialized to $p = 1$ and $q = 12$, the resulting expressions correspond to the PS-b-P(MMA_x-co-LMA_{1-x}) diblock copolymers studied by Ruzette et al. [113].

Within the BLCT, χ_s is positive, while χ'_h can formally be of either sign, depending on the signs of the three effective interaction energies ε_1, ε_2, and ε_3 (and on the composition x when one of the blocks is a random copolymer). We have found [114] a plethora of sets $\{\varepsilon_i\}$ for which the enthalpic portion χ'_h of the interaction parameter χ is positive for PS-PMMA ($n = 1$), negative for PS-PEMA, PS-PPMA, and PS-PBMA diblock copolymers ($n = 2$, 3, and 4, respectively), positive again for PS-PHMA, PS-POMA, and PS-PLMA melts ($n = 6$, 8, and 12, respectively), and, finally, negative for the block copolymer of styrene and a 53–47 weight percent random copolymer of methyl ($n = 1$) and lauryl methacrylates ($n = 12$) [i.e. for PS-b-P($MMA_{0.74}$-co-$LMA_{0.26}$)]. Systems with a positive χ'_h order on cooling, while systems with a negative χ'_h order on heating. The existence of numerous acceptable sets of energy parameters satisfies our primary goal of elucidating a molecular mechanism that explains the observed variation of the order–disorder transition type with the length of the alkyl side group in the *n*-alkyl methacrylates. This remarkable behavior is explained in terms of different interactions between chemically different united atom groups. The interpretation of this behavior does not require considering finite compressibility effects nor resorting to a sophisticated theory, although finite compressibility, nonrandom mixing effects, etc. must affect the experimental χ_{SANS} parameters, and, hence, the observed ODT transition temperatures.

Table 4 summarizes the χ'_h that are determined from Eqs. 48b and 49b for a typical set of $\{\varepsilon_i\}$ (with the constraints $\varepsilon_1 > 0$, $\varepsilon_2 < 0$, and $\varepsilon_3 > 0$), along with the calculated entropic portions χ_s and the order–disorder transition temperatures T_{ODT} for the eight PS-b-P*n*AMA systems studied experimentally [113]. The computed values of T_{ODT} for certain diblock copolymer melts turn out to be very sensitive to the value of χ_s that is substituted into the denominator of Eq. 47. A small variation in χ_s for these systems leads to a huge change in the computed T_{ODT}. Consequently, instead of a single value for T_{ODT}, Table 4 presents ranges of T_{ODT} that are obtained by ascribing an *uncertainty* of ±20% to the computed χ_s. Taking into account the obvious simplification inherent in the use of a lattice model and, in particular,

Table 4 Order–disorder transition temperature of polystyrene-b-poly(*n*-alkyl methacrylate) diblock copolymers and the polystyrene-b-poly($MMA_{0.74}$-co-$LMA_{0.26}$) copolymer system as determined from SANS experiments [113] and from BLCT calculations[a]

Sample	Theory			Experiment
	$\chi_s \times 10^2$	χ'_h[K]	T_{ODT}[K]	T_{ODT}[K]
28 K PS-b-PMMA	4.14	2.14	303 to ∞	< 373
50 K PS-b-PEMA	3.12	- 3.11	214–1501	> 463
110 K PS-b-PPMA	2.41	- 3.75	220–505	473
75 K PS-b-PBMA	1.88	- 1.67	0 to ∞	473
27 K PS-b-PHMA	1.18	6.78	163–184	< 373
23 K PS-b-POMA	0.759	17.8	327–346	< 373
19 K PS-b-PLMA	0.315	41.7	446–452	408
80 K PS-b-($MMA_{0.74}$-co-$LMA_{0.26}$)	1.95	- 2.12	248 to ∞	441

[a] The adjustable parameters of the theory are chosen as $\varepsilon_1 = 12.45$ K, $\varepsilon_2 =- 91.33$ K, and $\varepsilon_3 = 149.45$ K and are assumed to be independent of the deuterium content of the PS blocks

the modeling of monomer structures within the geometric constraints of a cubic lattice ($z = 6$), this choice of uncertainty is quite reasonable. Given the error bars for the computed χ_s, the agreement between theoretical and experimental transition temperatures T_{ODT} appear satisfactory for all but the PS-b-PHMA system where the computed T_{ODT} is probably too low.

10 Discussion

The lattice cluster theory (LCT) has been developed [12–22] to rectify several glaring deficiencies in the highly useful Flory–Huggins (FH) theory [3–7] for the thermodynamics of polymer systems. More specifically, the LCT describes the variation of thermodynamic properties of polymer systems with the monomer molecular structures of the constituents. This description is based on the systematic computation of the nonrandom mixing corrections to FH that arise from the presence of local correlations, which are present on length scales smaller than average monomer sizes and which are completely neglected in FH theory. The present review focuses on two simplifying limits of the LCT that clearly illustrate the manifold physical ramifications of this molecular theory.

The simplified lattice cluster theory (SLCT) [46–48] emerges from the LCT by neglecting contributions in the noncombinatorial portion of the specific free energy of mixing that depend on the excess free volume fraction and on the molecular weights. Thus, the SLCT coincides with the LCT in the high molecular weight, high pressure limit. The binary blend SLCT differs from the corresponding FH theory in two respects. Firstly, the SLCT predicts the presence of a temperature-independent *entropic* contribution χ_s to the effective interaction parameter χ_{SANS} that stems from the structural asymmetry between the two components of the blend and that is absent in FH theory. Importantly, the evaluation of χ_s in the SLCT proceeds by the trivial counting of the numbers of different types of united atom groups in single monomers of the two blend components. Comparisons with experiment indicate that the computed values of χ_s are quite good so long as χ_s is not too small or is negative. (Indeed, negative values of χ_s are rare for homopolymer blends.) The second relevant feature distinguishing the SLCT from FH theory is the dependence of the energetic portion χ_h' of the SLCT χ_{SANS} on the monomer structures of the two blend components and on the blend composition. While a composition-dependent term in χ_h' is introduced into several phenomenological treatments of polymer blends in a physically vague or ad hoc fashion, this term appears in the SLCT as a direct consequence of monomer structural asymmetry. Both χ_s and the composition-dependent portion of χ_h' vanish identically when the two blend species have identical structures. These two basic improvements beyond FH theory are responsible for the ability of the SLCT to predict several types of critical phase behavior whose explanation lies beyond the scope of FH theory [47, 48].

The SLCT has been applied to several binary homopolymer polyolefin blends that have been widely studied experimentally and that display behaviors that cannot be rationalized with FH theory. These applications illustrate how the molecular formulation of the SLCT aids in understanding nontrivial miscibility phenomena in these systems. For instance, the SLCT explains why hhPP is far more miscible with PEP than is PP and why blends of PIB with several other polyolefins tend to form lower critical solution temperature (LCST) miscibility patterns. Further comparisons of the SLCT with experiment will undoubtedly reveal limitations of the theory that result from the use of a lattice model, of oversimplistic united atom models, of monomer averaged interaction energies, etc., as well as limitations due to the fact that the SLCT only contains the leading monomer-structure-dependent contribution to the free energy. Thus, some significant improvements should emerge from the computation of higher order contributions to the LCT free energy [117].

When applied to systems containing statistical copolymers, the SLCT loses its enormous analytical and physical simplicity due to the greater complexity of these systems compared to binary homopolymer blends. The lack of mathematical simplicity in describing copolymer blends arises, in part, from the dependence of the free energy on the monomer sequence. Therefore, a further approximation is introduced into the SLCT to generate a theoretical approach that is simple and easy to use but is devoid of a serious deficiency of the extensions of FH theory to random copolymer systems, namely, the neglect of the temperature-independent portion χ_s of χ_{SANS}. The resulting theory is termed the basic lattice cluster theory (BLCT) to emphasize its relation to the LCT. Like the SLCT, the BLCT applies to binary blends in the high molecular weight, high pressure limit, but, in contrast to the SLCT, the BLCT ignores the monomer-structure-dependent contribution to the interaction term χ'_h. Thus, the neglect of this term is the only difference between the two theories, but this omitted term implies that the BLCT χ_{SANS} is independent of composition. Applications of the BLCT to several copolymer systems demonstrate that χ_s plays an essential role in explaining a wide range of thermodynamic behavior.

10.1 Improved Theories

The SLCT and the BLCT are suitable theoretical tools to illustrate and predict the influences of monomer molecular structure and nonrandom mixing effects on thermodynamic properties of many polymer systems, but the more general LCT should be used when the dual assumptions of high molecular weights and/or incompressibility are no longer applicable. For instance, the pressure dependence of χ_{SANS}, which is discussed in the article by Schwahn in this volume, can be investigated theoretically with the LCT. Although the LCT expressions for the free energy of mixing ΔF_{mix} and the χ_{SANS} parameter for a compressible system are far lengthier than those of the SLCT, the LCT corrections to ΔF_{mix} of FH theory are simple polynomials in the volume fractions of the constituents of the system [20, 21]. The volume fraction for the excess free volume is obtained numerically from the equation of state that is derived from the LCT free energy expression [25]. The LCT predicts a wider range of possible phase behaviors than the SLCT, such as ones with closed-loop phase diagrams or with miscibility windows [59, 118]. The description of compressible systems, however, necessitates the introduction of additional adjustable parameters into the theory and, consequently, depends on the availability of more extensive experimental data. For example, the equivalent of SLCT Eq. 9a for compressible binary blends contains three independent monomer averaged interaction energies ε_{11}, ε_{22}, and ε_{12} rather than the single exchange energy ε that appears in the SLCT Eq. 9a. While the self-interaction energies ε_{11} and ε_{22} are usually determined from fits to data for pure melts, the heterocontact energy ε_{12} can only be determined from fits to experimental data for the binary blend. Prior papers explain the technical details of these fits [25, 26].

As mentioned earlier, the SLCT has been derived for A_xB_{1-x}/C_yD_{1-y} copolymer systems [93], but the theory has not yet been applied to analyze experimental data. In contrast to the BLCT, the SLCT predicts that the energetic portion χ'_h of the χ_{SANS} parameter for copolymer blends depends on the blend composition and is sensitive to the monomer sequence distribution, i.e., differs for blends of alternating, diblock, and random copolymers. The SLCT expression for ΔF_{mix} of A_xB_{1-x}/C_yD_{1-y} systems has a form similar to Eq. 9a, but ΔF_{mix} for the copolymer system also contains a monomer-sequence-dependent contribution.

The remarkable ordering behavior in the series of PS-b-P*n*AMA diblock copolymer melts has been explained by lifting the assumption of monomer averaged interaction energies and by ascribing different interaction energies to chemically different united atom groups within a monomer. However, this generalization leads inevitably to an increase in the number of adjustable parameters in the theory. While it would be highly desirable to accumulate a set of united-atom-group-dependent interaction energies that could be used for a wide range of different systems, this worthwhile goal requires considerable analysis and the collection of a rather large body of experimental data.

The LCT and its more simplified descendants, the SLCT and BLCT, contain only the leading monomer-structure-dependent contributions from the cluster expansion. While this leading contribution has already provided a wealth of molecular-scale understanding concerning the miscibility and thermodynamic properties of binary polymer blends, the computation of the next order contributions [117] should improve the predictive powers of the theory to enable describing, for instance, the tacticity and sequence dependence of ΔF_{mix} in copolymer systems since the inclusion of these latter effects requires treating local correlations on the longer-length scales that have been considered so far.

Recently, we have extended [81] the Gibbs–DiMarzio theory of the glass transition in systems of semiflexible, interacting polymers composed of structured monomers. This extension is based on a reformulation of the LCT that enables the computation of the configurational entropy S_c that is a basic quantity in many theoretical treatments of the glass transition. This new theoretical approach provides a valuable tool for determining how monomer structure affects glass transition temperatures of polymer melts (J. Dudowicz, K.F. Freed, J.F. Douglas unpubl. results) by [119].

Lattice models have also been very popular in theoretical studies of proteins and, in particular, protein folding. Each amino acid is generally treated in these lattice models as a *united residue* entity occupying a single lattice site in the spirit of the original Flory–Huggins lattice model for polymers. The striking differences in miscibilities observed for the chemically very similar PP/PEP and hhPP/PEP blends engender some skepticism toward the use of these popular protein folding lattice models. Given the presence of specific hydrogen bonding interactions in proteins, it is questionable whether extended lattice models in which individual amino acids can occupy several neighboring lattice sites can offer as much insights as found for binary blends of synthetic polymers.

References

1. Thayer AM (1995) Chem Eng News 11:15
2. Subramanian PS, Chou KJ (1995) Trends Polym Sci 3:324
3. Flory PJ (1941) J Chem Phys 9:660
4. Flory PJ (1953) Principles of polymer chemistry. Cornell University Press, Ithaca, NY
5. Huggins ML (1941) J Chem Phys 9:440
6. Huggins ML (1942) J Chem Phys 46:151
7. Huggins ML (1943) Ann NY Acad Sci 44:431
8. Sanchez IC, Lacombe RH (1976) J Phys Chem 80:2352
9. Sanchez IC, Lacombe RH (1978) Macromolecules 11:1145
10. Sanchez IC (1983) Annu Rev Mater Sci 13:387
11. Panayiotou CG (1987) Macromolecules 20:861
12. Flory PJ (1970) Disc Faraday Soc 49:7
13. Flory PJ, Orwoll RA, Vrij A (1964) J Am Chem Soc 86:3507
14. Flory PJ, Orwoll RA, Vrij A (1964) J Am Chem Soc 86:3515
15. Dee GT, Walsh DJ (1988) Macromolecules 21:811
16. Dee GT, Walsh DJ (1988) Macromolecules 21:815
17. Freed KF, Dudowicz J (1995) Trends Polym Sci 3:248
18. Nemirovsky AM, Bawendi MG, Freed KF (1987) J Chem Phys 87:7272
19. Freed KF, Bawendi MG (1989) J Phys Chem 93:2194
20. Dudowicz J, Freed KF (1991) Macromolecules 24:5074
21. Dudowicz J, Freed MS, Freed KF (1991) Macromolecules 24:5096
22. Dudowicz J, Freed KF (1991) Macromolecules 24:5112
23. Janssen S, Schwahn D, Mortensen K, Springer T (1993) Macromolecules 26:5587
24. Hammouda B, Bauer BJ (1995) Macromolecules 28:4505
25. Dudowicz J, Freed KF (1995) Macromolecules 28:6625
26. Frielinghaus H, Schwahn D, Dudowicz J, Freed KF, Foreman KW (2001) J Chem Phys 114:5016
27. Graessley WW, Krishnamoorti R, Balsara NP, Fetters LJ, Lohse DJ, Schulz DN, Sissano JA (1993) Macromolecules 26:1137
28. Graessley WW, Krishnamoorti R, Balsara NP, Fetters LJ, Lohse DJ, Schulz DN, Sissano JA (1994) Macromolecules 27:2574
29. Krishnamoorti R, Graessley WW, Balsara NP, Lohse DJ (1994) Macromolecules 27:3073
30. Graessley WW, Krishnamoorti R, Balsara NP, Butera RJ, Fetters LJ, Lohse DJ, Schulz DN, Sissano JA (1994) Macromolecules 27:3896
31. Graessley WW, Krishnamoorti R, Reichart GC, Balsara NP, Fetters LJ, Lohse DJ (1995) Macromolecules 28:1260
32. Reichart GC, Graessley WW, Register RA, Krishnamoorti R, Lohse DJ (1997) Macromolecules 30:3036
33. Rajasekaran JJ, Curro JG, Honeycutt JD (1995) Macromolecules 28:6843
34. Rajasekaran JJ, Curro JG (1995) J Chem Soc Faraday Trans 91:2427
35. Curro JG (1994) Macromolecules 27:4665
36. Mayer JE, Mayer MG (1940) Statistical Mechanics. Wiley, New York
37. Mondello M, Grest GS, Garcia AR, Sibernagel BG (1996) J Chem Phys 105:5208
38. Smit B, Karaboni S, Siepmann JI (1995) J Chem Phys 102:2126
39. Siepmann JI, Martin MG, Mundy CJ, Klein ML (1997) Mol Phys 90:687
40. Allen W, Rowley RL (1997) J Chem Phys 106:10273
41. Buta D, Freed KF, Szleifer I (2000) J Chem Phys 112:6040

42. Bawendi MG, Freed KF (1988) J Chem Phys 88:2741
43. Buta D, Freed KF (2002) J Chem Phys 116:10959
44. Buta D, Freed KF (2004) J Chem Phys 120:6288
45. Schweizer KS, Curro JG (1997) Adv Chem Phys 98:1
46. Freed KF, Dudowicz J (1998) Macromolecules 31:6681
47. Dudowicz J, Freed KF, Douglas JF (2002) Phys Rev Lett 095503
48. Dudowicz J, Freed KF, Douglas JF (2002) J Chem Phys 116:9983
49. Foreman KW, Freed KF (1998) Adv Chem Phys 103:335
50. Nemirovsky AM, Dudowicz J, Freed KF (1992) Phys Rev A 45:7111
51. Freed KF, Dudowicz J (1996) Macromolecules 29:625
52. Dudowicz J, Freed KF (1997) Macromolecules 30:5506
53. Jinnai H, Hasegawa H, Hashimoto T, Han CC (1992) Macromolecules 25:6078
54. Sariban A, Binder K (1997) J Chem Phys 86:5859
55. Foreman KW, Freed KF, Ngola IM (1997) J Chem Phys 107:4688
56. Guggenheim EA (1994) Proc R Soc Lond A 183:203
57. Guggenheim EA (1952) Mixtures. Oxford University Press, Oxford
58. Foreman KW, Freed KF (1995) J Chem Phys 102:4663
59. Freed KF, Dudowicz J (1992) Theor Chim Acta 82:357
60. Krishnamoorti R, Graessley WW, Fetters LJ, Garner RT, Lohse DJ (1995) Macromolecules 28:1252
61. Patterson D (1969) Macromolecules 2:672
62. de Gennes PG (1979) Scaling Concepts in Polymer Physics. Cornell University Press, Ithaca, NY
63. de Gennes PG (1977) J Phys Lett 38:L441
64. Binder K (1983) J Chem Phys 79:6387
65. Dudowicz J, Lifschitz M, Freed KF, Douglas JF (1993) J Chem Phys 99:4804
66. Freed KF (1987) Renormalization Theory of Macromolecules. Wiley, New York
67. Binder K (1984) Phys Rev A 29:341
68. Anisimov MA, Kisielev SB, Sengers JV, Tang S (1992) Physica A 188:487
69. Levelt-Sengers JMH, Sengers JV (1981) In: Raveche HJ (ed) Perspectives in statistical physics. North Holland, Amsterdam
70. Holyst R, Vilgis TA (1993) J Chem Phys 99:4835
71. Gehlsen MD, Rosendale JH, Bates FS, Wignall GD, Hansen L, Almdal K (1992) Phys Rev Lett 68:2452
72. Müller M, Binder K (1995) Macromolecules 28:1825
73. Escobedo FA, de Pablo JJ (1999) Macromolecules 32:900
74. Han CC, Bauer BJ, Clark JC, Muroga Y, Matsushita Y, Okada M, Tran-cong Q, Chang T, Sanchez IC (1988) Polymer 29:2002
75. Han CC, Okada M, Muroga Y, Bauer BJ, Tran-cong Q (1986) Pol Eng Sci 26:1208
76. Choi S, Lin X, Briber RM (1988) J Polym Sci B 36:1
77. Briber RM, Bauer BJ, Hammouda B (1994) J Chem Phys 101:2592
78. Flory PJ (1956) Proc R Soc Lond A 234:60
79. Bawendi BG, Freed KF (1987) J Chem Phys 87:3720
80. Foreman KW, Freed KF (1997) Macromolecules 30:7279
81. Freed KF (2003) J Chem Phys 119:5730
82. Freed KF, Dudowicz J, Foreman KW (1998) J Chem Phys 108:7881
83. Freed KF, Dudowicz J unpublished results
84. Paul DR, Barlow JW (1984) Polymer 25:487
85. Roe RJ, Rigby D (1987) Adv Polym Sci 82:103
86. Huh W, Karasz FE (1992) Macromolecules 25:1057

87. Sakurai S, Hasegawa H, Hashimoto T, Hargis IG, Aggarwall SL, Han CC (1990) Macromolecules 23:451
88. ten Brinke G, Karasz FE, MacKinight WJ (1983) Macromolecules 16:1827
89. Kambour RP, Bendler JT, Bopp RC (1983) Macromolecules 16:753
90. Balazs AC, Sanchez IC, Epstein I, Karasz FE, MacKnight WJ (1985) Macromolecules 18:2188
91. Cantow HJ, Schulz O (1986) Polym Bull 15:449
92. Kohl PR, Seifert AM, Hellemann GP (1990) J Polym Sci B 28:1309
93. Dudowicz J, Freed KF (2001) Polish J Chem 75:527
94. Dudowicz J, Freed KF (1996) Macromolecules 29:7826
95. Dudowicz J, Freed KF (2000) Macromolecules 33:3467
96. Delfolie C, Dickinson LC, Freed KF, Dudowicz J, MacKnight WJ (1999) Macromolecules 32:7781
97. Scheffold F, Eiser E, Budkowski A, Steiner U, Klein J, Fetters LJ (1996) J Chem Phys 104:8786
98. Rowlinson JS, Swinton FL (1982) Liquids and liquid mixtures. Butterworths, London
99. Bates FS, Fetters LJ, Wignall GD (1988) Macromolecules 21:1086
100. Schweizer KS, David EF, Singh C, Curro JG, Rajasekaran JJ (1995) Macromolecules 28:1528
101. Schweizer KS, Singh C (1995) Macromolecules 28:2063
102. Singh C, Schweizer KS (1997) Macromolecules 30:1490
103. Rhee J, Crist B (1992) Polym Mater Sci Eng 67:209
104. Budkowski A, Klein J, Eiser E, Steiner U, Fetters LJ (1993) Macromolecules 26:3858
105. Weimann PA, Jones TD, Hillmyer MA, Bates FS, Londono JD, Melnichenko Y, Wignall GD (1997) Macromolecules 30:3650
106. Bates FS, Wignall GD, Koehler WC (1985) Phys Rev Lett 55:2425
107. Bates FS, Dierker SB, Wignall GD (1986) Macromolecules 19:1938
108. Dudowicz J, Freed KF (2000) Macromolecules 33:9777
109. Schwahn D, Willner L (2002) Macromolecules 35:239
110. Dudowicz J, Freed KF (1993) Macromolecules 26:213
111. Russell TP, Karis TE, Galot Y, Mayes AM (1994) Nature 368:729
112. Karis TE, Russell TP, Galot Y, Mayes AM (1995) Macromolecules 28:1129
113. Ruzette AVG, Banarjee P, Mayes AM, Pollard M, Russell TP, Jerome R, Slawecki T, Hjelm R, Thiyagarajan P (1998) Macromolecules 31:8509
114. Dudowicz J, Freed KF (2000) Macromolecules 33:5292
115. Leibler L (1980) Macromolecules 13:1602
116. Bates FS, Rosedale JH, Fredrickson GH (1990) J Chem Phys 92:6255
117. Brazhnik O, Freed KF (1996) J Chem Phys 105:837
118. Dudowicz J, Freed KF (1996) Macromolecules 29:8960
119. Dudowicz J, Freed KF, Douglas JF (2005) J Phys Chem B (in press)

Adv Polym Sci (2005) 183: 127–173
DOI 10.1007/b135884

Published online: 26 July 2005

Effect of Shear Flow on Polymer Blends

Nigel Clarke

Department of Chemistry, University of Durham, via Cintia, Durham DH1 3LE, UK
nigel.clarke@durham.ac.uk

Abstract An overview of the various theoretical approaches to understanding the effect of shear flow on polymer blends is presented. Two differing models have emerged; one proposes the addition of a shear-dependent term directly into the free energy, whereas the other considers how shear modifies the equation of motion for concentration fluctuations. As discussed in detail, despite these differences, the models share much in common, and some degree of convergence between the two approaches has recently emerged. Although the focus is on modelling, a brief review of the varied range of experimental observations, highlighting the challenges that face theoreticians, is also given.

1 Introduction

The consequences of flow on the structure of fluid mixtures have received considerable attention. Interests in the subject stem not only from its industrial relevance – many processing technologies involve the flow of mixtures, but also from its importance from a fundamental point of view – understanding and predicting behaviour provides a challenge and a test for non-equilibrium thermodynamics. A significant proportion of industrial mixtures can be considered as either polymer solutions or blends, the latter in particular will form the basis for this review. Since the effects of shear on polymer solutions have been the subject of at least two extensive reviews, we will focus here attention on polymer blends, and the unique behaviour associated with them.

Firstly, it is useful to define the specific nature of flow that is of relevance. The two principal types of flow are shear and extensional, both of which play an important role during processing. Much attention has been devoted to the former since it is more straightforward to establish a steady-state shear flow than a steady-state extensional flow, and, in general, the response of single-component polymeric liquids to shear flow is better understood than their response to extensional flow. The three most common geometries for studying polymer systems subjected to flow are concentric cylinders, cone and plate, and parallel plates (see Fig. 1). The shear flow is characterised by the shear rate, $\dot{\gamma}$, such that if we denote x as the flow direction, the velocity is given by $v = \dot{\gamma} y \vec{x}$, where y denotes the shear gradient direction. Concentric cylinders are useful for low viscosity solutions, whereas the cone and plate and parallel plate set-ups are utilised for more viscous solutions and blends. In the cone and plate geometry the shear rate is constant across the radius, whereas in the parallel plate geometry the shear rate increases linearly with distance from the centre of the plates.

Interest in the response of polymer blends subjected to shear ranges from shear-induced homogenisation of incompatible blends to shear -induced phase separation of miscible blends. Shear has also been used to deform existing microstructures in immiscible polymer mixtures. However, this review article is mostly concerned with *partially miscible* polymer blends; i.e., blends whose miscibility changes with temperature. Such blends are categorised as exhibiting either lower critical solution temperature (LCST) or upper critical solution temperature (UCST) behaviour. The former phase separate upon heating, whilst the latter phase separate upon cooling. The phase separation

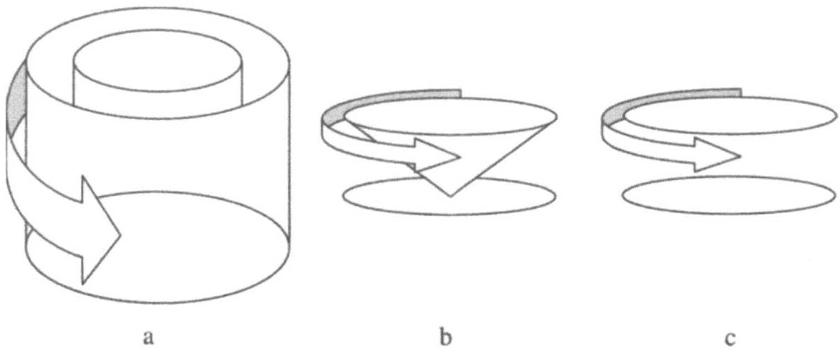

Fig. 1 Geometries used for applying shear flows. (**a**) Concentric cylinders (Couette flow) – either the inner or outer cylinder is rotated, (**b**) Cone and plate, and (**c**) parallel plates

temperature, often referred to as the cloud point (CP), since transparent miscible mixtures become cloudy when they start to phase separate, depends upon the chemical nature of the two components, the molecular weights and polydispersity of each and the composition.

2 Experimental Studies

It is useful to first discuss some of the unusual observations that continue to motivate interest in this subject. Perhaps the simplest and most important question to address in partially miscible polymer blends is whether shear affects the phase boundary. In particular, it is desirable to know whether shearing increases or decreases the range over which a blend is compatible. A number of experimental techniques have been utilised to help answer this question. An early study by Mazich and Carr [1] inferred shifts in the phase boundary of polystyrene/polyvinylmethylether (PS/PVME). This is an LCST blend in quiescent conditions, and is ideal for shear flow studies, since the phase boundary is generally well above the glass transition temperature of the blend. The location of the phase boundary was assumed to be indicated by an abrupt change in the temperature dependence of the viscosity. For two different mixtures of PS/PVME (see blends and 1 and 2 in Table 1 for details), it was found that the boundary was shifted to higher temperatures by 2–7 K, across a range of compositions, indicating shear-induced mixing. Rather than using the shear rate as the fixed parameter, they chose to fix the first normal stress, arguing that this leads to constant stretching of the chain molecules at different temperatures. In many instances it is more convenient to use shear rate as the control parameter; in such a case it is necessary to interpret results in view of the changing viscosity with temperature. In other words, at a fixed shear rate the shear-induced deformation of individual chains will be strongly temperature dependent.

Katsaros et al. [7, 8] studied PS/PVME blends and determined shifts in the phase boundary using optical clarity. The disadvantage of this technique is that although the shear may prevent large-scale morphologies from developing, the blend may remain phase separated at a much smaller length scale than that of the wavelength of light, so that it appears transparent. More direct evidence was gathered by the same group using fluorescence studies. Anthracene labelled polystyrene (PS*) was added in trace quantities to a PS/PVME blend. The PS* fluorescence is quenched by the presence of the ether group in PVME, so that molecular level mixing or de-mixing can be determined by the degree of fluorescence. In the preliminary work of Mani et al. [2] (see blend 3 in Table 1 for details), blends were held at 3 K above their cloud points, so that the mixtures were phase separated. They were

Table 1 Details of some of the partially miscible blends used for studying the effect of shear on phase separation

Blend	Components	Molecular weight/kg mol^{-1}	Reference
1	Polystyrene	93	[1]
	Poly(vinylmethylether)	102	
2	Polystyrene	411	[1]
	Poly(vinylmethylether)	62	
3	Polystyrene	348	[2]
	Poly(vinylmethylether)	48	
4	Polystyrene	259	[3]
	Poly(vinylmethylether)	85	
5	poly(ethylene-co-vinyl acetate)	100	[4]
	solution-chlorinated polyethylene	332	
6	Poly(butyl acrylate)	75.5	[4]
	Solution-chlorinated polyethylene	236	
7	Polystyrene	302	[4]
	Poly(vinylmethylether)	69.7	
8	Polystyrene	330	[5]
	Poly(vinylmethylether)	89	
9	Polystyrene	65	[5]
	Poly(vinylmethylether)	89	
10	Poly (styrene-acrlyonitrile)	190	[11]
	Poly ε-caprolactone	95	

then sheared for different amounts of time and rapidly quenched below the glass transition temperature of the blend to freeze-in the shear-induced structures. Above a critical shear rate it was found that the blends became mixed; however a reduction of the critical shear rate with shearing time was also observed. This phenomenon highlights the importance of time-dependent effects: even if it is in some sense thermodynamically preferable to remix, the system has to evolve from one state to another. It is therefore essential to ensure that a steady state has been achieved. In such an experiment, it is also to be expected that the time the blend is held above the CP prior to shearing will also play an important role: the longer the blend has been allowed to phase separate, the larger the phase separated structures will be. At the very least it is to be expected that the size of the microstructures will affect the time taken for a steady state to be reached.

Larbi et al. [3] performed in-situ fluorescence studies (see blend 4 in Table 1 for details), using a modified rheometer. Samples were heated slowly at constant shear rates, and again phase separation was suppressed to higher temperatures. For a range of different compositions the shift, ΔT, in the demixing temperature with respect to the quiescent phase boundary, T_s, was well described by the empirical relation,

$$\Delta T/T_s = (0.015 \pm 0.002)\dot{\gamma}^{0.59\pm0.04} \tag{1}$$

Further fluorescence studies by Mani et al. [9] on the same blend discovered that de-mixing could be induced at as much as 40 K below the quiescent boundary at higher shear rates.

Hindawi et al. [4] utilised small-angle light scattering (SALS) to probe shear-induced changes in blends of poly(ethylene-co-vinyl acetate)/solution-chlorinated polyethylene (EVA/SCPE), poly(butyl acrylate)/SCPE (PBA/SCPE) and PS/PVME (see blends 5–7 in Table 1 for details). The advantage of SALS over fluorescence and rheological techniques is that it is particularly sensitive to small fluctuations in local concentration. It has been widely used to study phase separation in the quiescent state, since it allows not only phase boundaries to be determined but also the dynamics of the phase separation process to be followed. When a homogeneous blend is quenched into the unstable region of its phase diagram it phase separates by the mechanism of spinodal decomposition. Ever present fluctuations in the concentration start to grow rather than decay, and a dominant length-scale emerges. This length scale is, in many cases, comparable to the wavelength of light, and so the blend, rather than being transparent, becomes turbid due to scattering. Since the extent of scattering increases with time, SALS provides a quantitative method for probing phase separation. One disadvantage of the SALS technique is that it is not sensitive to phase separation at sub-micron length-scales, and so it is necessary to be aware that a lack of scattering does not always imply a lack of phase separation. In reference [4] the blends were sheared at a constant rate using a plate-plate shear cell and then slowly heated (both EVA/SCPE and PBA/SCPE exhibit quiescent LCST phase behaviour); the onset of phase separation was indicated by an increase in the light scattered from the sample at small scattering angles. A remarkable variety of behaviour was observed in the blends; in EVA/SCPE an increase in scattering occurred at lower temperatures, with respect to the quiescent boundary, at low shear rates, indicating an enhancement of the two-phase region, but at higher shear rates the behaviour was reversed with shear apparently suppressing phase separation. The maximum change in the phase boundary, – 5 °C, occurred at a shear rate of $\sim$1.5 s^{-1}, This behaviour was observed for a wide range of compositions. Similar trends were also observed in the PBA/SCPE and PS/PVME blends, although the maximum shifts, at a shear rate of $\sim$3–4 s^{-1}, were as high as – 10 and – 30 °C, respectively. It has been suggested that the changes in the boundary may be an artefact due to viscous heating of the sample. It was estimated that such heating would at most change the sample temperature by $\sim$1 °C, which is too small to account for the observed shifts. In order to provide further evidence for the shear-induced de-mixing at low shear rates and shear-induced mixing at higher shear rates, sheared blends were quenched below the glass transition temperature (T_g and heat flow scans were performed using differential scanning calorimetry). Phase separated blends show two T_gs, which for a blend with the two phases composed of pure A and pure B, will correspond to the glass transition temperatures of the two pure components. Thus the values of the measured glass transition temperatures also provide information on the degree of phase separation. The results from the SALS were indeed confirmed, suggesting that it was not the case that phase separation was occurring over different length scales inaccessible to light scattering. Fernandez et al. [5] followed this work with studies of PS/PVME blends and found that at a constant shear rate as the temperature was increased phase separation occurred as much as 15 °C below the quiescent phase boundary, depending on the shear rate. As the temperature was further increased remixing was observed and finally de-mixing at even higher temperatures. This behaviour, shown in Fig. 2, was interpreted as being the result of an island of miscibility in the shear rate – temperature "phase diagram".

Madbouly et al. [10] carried out a systematic study on blends of poly styrene-acrlyonitrile (SAN) and poly methyl methacrylate (PMMA) with varying PMMA molecular weights ranging from 7 to 396 kg mol^{-1}. Significantly, when compared to other results, only shear-induced mixing was observed for all samples. Unsurprisingly, the shear effect was found to be composition and molecular weight dependent. As the shear range was increased, the shift in the cloud point first increased monotonically with shear rate and then become almost constant regardless of the applied shear rate values. A maximum elevation in the normalized shift of the cloud points was found

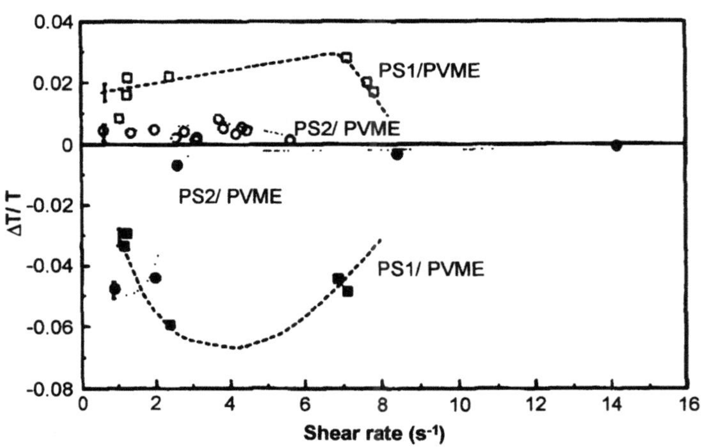

Fig. 2 Normalised shift in the cloud point curve as a function of shear rate for blends of polystyrene and poly(vinyl methyl ether). The blend labelled PS1/PVME (*squares*) and PS2/PVME (*circles*) correspond to blend 8 and 9 respectively in Table 1. In each case, two cloud points were observed as the blends were simultaneously sheared and heated. Reproduced with permission from reference [5]

when the viscosity ratio is close to unity; a result which is difficult to understand in the framework of the current theories discussed below that suggest that shifts in phase boundaries should be greater when the viscosity mismatch is greater.

More recently [11], for a blend of SAN and poly ε-caprolactone (see blend 10 in Table 1 for details), no shear-induced de-mixing was observed. However, only one measurement at one shear rate and one temperature within the one-phase region (10 °C below the phase boundary) was reported, and since shifts are often by smaller amounts this result is far from conclusive.

The dynamics of shear-induced phase separation during the early stages of the process have been studied by Gerard et al. [12–14] immediately following the imposition of shear using SALS. It was found that the growth rate of concentration fluctuations was strongly dependent on the shear rate, and qualitative agreement with some of the theoretical predictions to be discussed below was found.

3
Theoretical Descriptions

In order to understand how shear flow affects the phase behaviour of polymer blends, it is useful to consider first how homogeneous polymeric systems respond to an imposed shear flow. This is a subject that has received considerable attention, and many of the details are unnecessary for this review; however, a basic understanding of polymer rheology [15] is essential, and so we present a brief overview of the most relevant ideas. When a polymer solution or blend (or indeed any viscoelastic medium) is subjected to a steady shear, its response can be characterised by the viscosity,

$$\eta = \frac{\sigma_{xy}}{\dot{\gamma}} \tag{2}$$

where σ_{xy} is the shear stress, and the first normal stress coefficient,

$$\Psi_1 = \frac{N_1}{\dot{\gamma}^2} = \frac{\sigma_{xx} - \sigma_{yy}}{\dot{\gamma}^2} \tag{3}$$

where N_1 is the first normal stress difference. Many polymeric systems also exhibit second normal stress differences, $N_2 = \sigma_{yy} - \sigma_{zz}$, but these are usually small compared to N_1, and their consequences have so far been neglected in most theories related to polymer blends. It is well known that polymers have unusual properties in the liquid state, the clearest example being the observation that the zero shear viscosity, η_0, the viscosity in the limit $\dot{\gamma} \to 0$, scales with molecular weight as $\eta_0 \propto M^{3.4}$ for polymers above a critical molecular weight. At low shear rates polymer solutions and blends exhibit so-called Newtonian behaviour, that is the both the viscosity and the first normal stress coefficient are independent of the shear rate. At higher shear rates, shear thinning is observed.

The behaviour of short polymers in the melt is reasonably well described by the Rouse model, originally proposed to explain the dynamics of dilute polymer solutions, in which the motion of a polymer is represented by the Brownian motion of beads connected by Hookian springs with no other interactions. However, from Table 1, it is clear that in order for shear to affect the stability of blends, the components generally have reasonably high molecular weights. The foundation for modern theories of the flow behaviour of such polymers was proposed by deGennes [16, 17], who, by analogy with the theory of rubber elasticity for crosslinked polymers, proposed the tube model. This was then explored rigorously by Doi and Edwards [18–21], and has been successfully applied to many aspects of the dynamics of polymers. The tube model is an elegant attempt to overcome the mathematical difficulty of describing the consequences of the interaction of all the surrounding polymers on the motion of any given polymer. The effect of the neighbours is modelled by a tube with a given diameter around a polymer, and it is assumed that the only motion is along the axis of the tube – chains cannot move laterally through tubes. This motion has been called reptation, and the chains are described as entangled. The diameter of the tube is the average distance between effective crosslinks. These "crosslinks", or entanglements, only prevent the chains from moving through each other, not sliding around each other. An important property of an entangled chain is its relaxation function, this describes the fraction of the chain that at time t still resides within the tube surrounding the chain at some earlier time t'. To a reasonable approximation the relaxation function is single exponential with a time constant, τ_R, known as the reptation time. The tube model predicts that τ_R scales with molecular weight as M^3. Following a small deformation the stress is assumed to be proportional to the number of initial entanglements that have not relaxed, with a proportionality constant, the plateau modulus G_0, given by the theory of rubber elasticity for the modulus of a crosslinked rubber, with the molecular weight between crosslinks replaced by the molecular weight between entanglements. Hence, if a polymeric system is subjected to a small step strain the stress relaxes as,

$$G(t-t') = G_0 \exp\left\{-\frac{t-t'}{\tau_R}\right\} \tag{4}$$

During steady state shearing, stress bearing entanglements are continually created and relaxed, and the stress tensor is given by,

$$\sigma = \int_{-\infty}^{t} \frac{\partial G(t-t')}{\partial t} C^{-1}(t,t')\mathrm{d}t' \tag{5}$$

where C^{-1} is known as the Finger tensor, which for shear flow is given by,

$$C^{-1}(t,t') = \begin{pmatrix} 1+\dot{\gamma}^2\left(t-t'\right)^2 & \dot{\gamma}\left(t-t'\right) & 0 \\ \dot{\gamma}\left(t-t'\right) & 1 & 0 \\ 0 & 0 & 1 \end{pmatrix} \tag{6}$$

This model, often referred to as the upper convective Maxwell model, is weakly non-linear in that it predicts a first normal stress, but no shear thinning effects, i.e, the shear stress increases linearly with shear rate so that the viscosity is independent of shear rate. Combining Eqs. 2, 4, 5 and 6, we see that the tube model predicts the viscosity to be,

$$\eta = G_0\tau_{\mathrm{R}} \propto M^3 \tag{7}$$

and the first normal stress coefficient to be,

$$\Psi_1 = 2G_0\tau_{\mathrm{R}}^2\dot{\gamma}^2 \tag{8}$$

The molecular weight dependence of the viscosity arises because the plateau modulus is independent of chain length. The discrepancy between the predicted and observed scaling of the viscosity has been the subject of intensive research. One of the principal assumptions of the original tube model is that the entanglements that form the surrounding tube are fixed. Since these entanglements arise from interactions with other chains which are themselves continuously moving, much effort has been put into developing models in which not only a polymer reptates from a tube but also the tube itself changes with time. A key feature of the chains in a binary polymer blend is how the dynamics depends on the local concentration. We will consider one of the conceptually simpler approaches to this problem in a later section.

3.1
The Thermodynamic Approach

Since shear induces apparent shifts in the phase boundary of blends, it is natural to attempt an explanation based on equilibrium thermodynamics. Horst and Wolf [22] extended a phenomenological model devised for polymer solutions proposed by Wolf [23]. The affect of shear was accounted for by adding a shear-dependent term, E_{s}, to the Gibbs free energy of mixing, F_{mix},

$$F_{\dot{\gamma}} = F_{\mathrm{mix}} + E_{\mathrm{s}} \tag{9}$$

The usual Flory-Huggins form was adopted for the zero-shear term,

$$\frac{F_{\mathrm{mix}}}{k_{\mathrm{B}}T} = \frac{\phi_{\mathrm{A}}}{N_{\mathrm{A}}}\ln\phi_{\mathrm{A}} + \frac{\phi_{\mathrm{B}}}{N_{\mathrm{B}}}\ln\phi_{\mathrm{B}} + \chi\phi_{\mathrm{A}}\phi_{\mathrm{B}} \tag{10}$$

where ϕ_{A} and ϕ_{B} are the volume fractions of A and B which have degrees of polymerisation N_{A} and N_{B}, respectively. The Flory-Huggins parameter, χ, is a measure of interactions between components. For convenience incompressibility, such that $\phi_A + \phi_B = 1$, is often assumed.

The stored energy term was assumed to be equal to the first normal stress difference, and to be given by the general expression,

$$E_s = \Psi_1 \dot{\gamma}^2 = (x_A V_A + x_B V_B) \langle J_e^0 \rangle \langle \eta \rangle^2 \dot{\gamma}^2 \, |\langle \eta \rangle \dot{\gamma}|^{-2d^*} \quad (11)$$

where $x_{A,B}$ and $V_{A,B}$ are the mole fractions and molar volumes of the polymers A and B, so that the first term in brackets represents the average molar volume. J_e^0, the steady-state shear compliance, is measured in experiments in which a steady shear stress is applied and the resultant shear rate measured, as opposed to determining the stress required to maintain a given shear rate. In general, J_e^0 is related to the plateau modulus by,

$$J_e^0 \approx 1/G_0 \quad (12)$$

The angular brackets denote that those parameters of Eq. 11 are measured for the homogeneous state. Finally, d^* is given by,

$$d^* =- \frac{\partial \ln \langle \eta \rangle}{\partial \ln \dot{\gamma}} \quad (13)$$

The physical basis of Eq. 11 is that the energy of the chains is increased during shear, due to the entropy of a stretched chain being less than that of free chain. Since chains with different chemical structures and molecular weights will respond differently to shear, the energy changes will depend on the concentration.

Equation 11 can be broken down into two factors. In the Newtonian regime the first normal stress for a polymer melt can be rewritten using Eqs. 8 and 12 as,

$$N_1 = \Psi_1 \dot{\gamma}^2 = 2G_0 \tau_R^2 \dot{\gamma}^2 = 2J_e^0 \eta^2 \dot{\gamma}^2 \quad (14)$$

It is important to note that the latter expression is completely general and applies independently of any model. However, as already noted, polymer systems become non-Newtonian at a critical shear rate, and shear-thinning is observed. An approximate relation gives the shear rate at which this occurs as,

$$\dot{\gamma} \tau_R \approx 1 \quad (15)$$

so that as the relaxation time, and hence viscosity, increases the range of shear rates at which Newtonian flow occurs decreases. At higher shear rates a power law relationship between N_1 and $\dot{\gamma}$ is found, hence Horst and Wolf proposed the phenomenological relationship leading to Eq. 13. In order to proceed, it is necessary to know how d^* varies with shear rate. An early version of the tube model for non-linear flows proposed by Graessley [24] was utilised. Shear thinning can be explained within the reptation framework by considering that at higher shear rates the number of entanglements experienced by a chain decreases because the contact time between two molecules decreases below the time required for an entanglement to form. Consequently, the viscosity and the first normal stress decrease as the number of entanglements decrease. Just as importantly it is necessary to consider how J_e^0 and η depend on the concentration of the components. Within the context of the original tube model the blending law is straightforward; since the entanglements are fixed, each polymer behaves independently of its surroundings, and the viscosity and first normal stress are simply the weighted sum, by the weight fractions of the two components, w_A and w_B, of the respective pure component values. However, Horst and Wolf made use of a different blending rule suggested by Schuch [25]. Rather than the stress relaxation functions being additive, it was assumed that the viscosity and compliance of a blend is given by that which would be appropriate for a melt with a molecular weight equivalent to the weight average molecular weight of the blend. This leads to,

$$\langle \eta_0 \rangle^{1/3.4} = w_A \eta_{0A}^{1/3.4} + w_B \eta_{0B}^{1/3.4} \tag{16}$$

and,

$$\langle J_e^0 \rangle = \left(w_A \eta_{0A}^{4.4/3.4} J_{eA}^0 + w_B \eta_{0B}^{4.4/3.4} J_{eB}^0 \right) / \eta_0^{4.4/3.4} \tag{17}$$

One of the key features of this and other theories is that there exists an asymmetry in the molecular response of each component to shear. Horst and Wolf determined shear dependent phase diagrams by applying the usual laws of thermodynamics to the total free energy. In reference [22] they determined shifts in the phase boundary using model parameters. They predicted that at low shear rates, shear-induced mixing should occur; followed by an inversion at higher shear rates to shear-induced de-mixing, followed by a further inversion at even higher shear rates to shear-induced mixing again. That two inversions occur is due to the fact that there are two different shear rates at which non-Newtonian flow behaviour occurs. The shear rate of the first inversion is determined by the more viscous component, whilst the second inversion is due to the onset of shear-thinning in the less viscous component. This argument is backed-up by the behaviour of polymer solutions [23], in which only the first inversion occurs, and there is only one critical shear rate. These comments only explain the existence of complex behaviour. However, the form of the mixing rules used makes it difficult to understand qualitatively the underlying physical mechanisms beyond the explanation given above.

It is worth highlighting the necessity of a non-linear mixing rule such as given by Eqs. 16 and 17. This is most easily done by considering the consequences of utilising a linear mixing rule, as would be predicted by the original tube model; in which the blend rheology is linearly dependent on the volume fraction, e.g.,

$$\eta_{Blend} = \phi_A \eta_A + \phi_B \eta_B \; ; \quad \Psi_{Blend} = \phi_A \Psi_A + \phi_B \Psi_B \tag{18}$$

This would result in no predicted perturbation due to shear. This is clear if we consider the thermodynamic rule for coexistence between two phases that the chemical potential be equal in each phase. For the approach of Horst and Wolf, the contribution to the chemical potential from the stored energy, given by,

$$\mu_{\mathrm{A}}^{\mathrm{shear}} = E_{\mathrm{S}} + \left(1 - \phi_{\mathrm{A}}\right)\frac{\partial E_{\mathrm{S}}}{\partial \phi_{\mathrm{A}}} \tag{19}$$

is simply a constant, since E_{S} would be linearly dependent on ϕ_{A} and will therefore not affect the coexistence conditions. Whether a more complex, and physically realistic, mixing rule leads to shear-induced mixing or de-mixing is more easily argued by considering the spinodal curve rather than the coexistence curve. The spinodal separates the regions in which the homogeneous state is metastable and unstable, and is related to the free energy by,

$$\frac{\partial^2 F}{\partial \phi^2} = 0 \tag{20}$$

If $\partial^2 F/\partial \phi^2 < 0$ the mixed state of the blend is unstable, so if the contribution that the stored energy makes to the spinodal is such that the curvature is negative than shear-induced de-mixing will result, whereas if the curvature is positive then shear-induced mixing occurs.

Fernandez et al. [5] applied the HW model, to explain the observed existence of an immiscibility gap, with limited success. As a consequence of this Horst and Wolf [26, 27] later modified their model to include important temperature dependencies of the rheological parameters. This lead to some complex predictions, most notably the possibility of closed loop miscibility gaps, as can be seen in Fig. 3. At certain shear rates, an LCST blend which is miscible at low temperatures, become immiscible far below the quiescent phase boundary as heated, but then become miscible again at a temperature still below quiescent phase boundary, finally becoming immiscible at some temperature above the quiescent phase boundary, which is in qualitative agreement with the findings of reference [5].

There has been a steady output modifying and further applying the theory. Horst and Wolf [28] calculated the influence on blends exhibiting UCST behaviour. Their predictions were similar to those for an LCST blend. As the shear rate is increased the blend experiences shear-induced mixing, then de-mixing, followed by the effect becoming negligible at higher shear rates. Miscibility gaps at certain shear rates, at temperatures above the quiescent phase boundary were also found. Krause and Wolf [29, 30] extended the model to include ternary blends of immiscible A and B polymers dissolved in a common solvent, and showed that their predictions compared favourably with experimental observations. Phase diagrams of sheared blends of homopolymer A and a random A-B copolymer [31] and ternary blends of homopolymer A, homopolymer B and a copolymeric compatibiliser have also

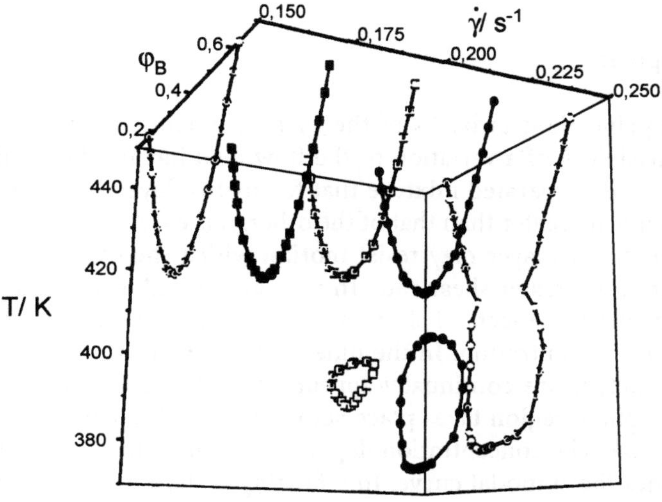

Fig. 3 Phase diagram for a model blend with components having molecular weights of 75 kg mol^{-1} and 200 kg mol^{-1}, calculated according to theory of Horst and Wolf. The existence of closed miscibility gaps above a critical shear rate is apparent. Reproduced with permission from reference [27]

been calculated [32, 33]. Sun et al. [34] have recently predicted the combined effects of shear and pressure changes on miscibility within the framework of the HW model.

Chopra et al. [35] have attempted to develop a microscopic basis for the methodology adopted phenomenologically by HW. They considered how flow affects the configurational choices, and thus the entropy, of a sheared polymer chain. It was proposed that the additional contribution to the free energy could be determined within the framework of the Flory-Huggins lattice theory, by weighting the probability that a monomer would occupy a given lattice space with respect to the previous monomer (on the same chain) in the direction of the flow. At higher shear rates, lattice spaces in the direction of flow become more heavily weighted. However a link between polymer dynamics, either for single chains or entangled chains was not established, and for calculation purposes, the authors resorted to utilising the same phenomenological ideas as HW. We also note that Jou et al. have discussed in detail for polymer solutions a justification based on extended irreversible thermodynamics (EIT). However, since the ideas have already been the subject of extensive reviews [36, 37], no further details will be included here. We note however that the EIT concept has recently been adapted for polymer blends [38], but since this work shares much in common with the dynamic approach introduced in the next few sections, further discussion will be deferred.

3.2 Dynamic Approach

One of the principle drawbacks of the thermodynamic approach is that, to date, the possibility of a variation of the flow field has not been allowed for. Consider a phase separated mixture that forms two layers. If the viscosity of one layer is much greater than that of the other, it is easy to imagine that when sheared, the viscous layer may resist motion whilst the other layer will flow at a compensatory faster shear rate. In partially miscible systems, it is common to consider the effects of shear by considering the behaviour of a local fluctuation in concentration. In the quiescent state within the one-phase region, such fluctuations continually appear, but will then decay. Once a jump into the two phase region takes place some of these fluctuations become unstable and grow. The concentration dependent temperature at which this first occurs defines the spinodal curve. In a flowing system, the existence of local fluctuations, and hence gradients, in concentration clearly results in gradients in the stress and the possibility of local deviations from the mean of the flow field itself. This in turn results in an additional factor that affects the stability of concentration fluctuations and hence the requirement for phase separation.

3.2.1 The Cahn–Hilliard Theory

It is essential to first briefly review the model for the dynamics of concentration fluctuations in the absence of shear, first derived by Cahn and Hilliard [39] for binary alloys, later modified by Cook [40] to include thermal noise, and more recently adapted by deGennes [41] and Pincus [42] for describing phase separation in polymer blends.

Since the total concentration of each polymer is conserved, then the continuity equation must be obeyed,

$$\frac{\partial \phi_A(\boldsymbol{r},t)}{\partial t} + \nabla \cdot \left[\boldsymbol{v}_A(\boldsymbol{r},t)\phi_A(\boldsymbol{r},t) \right] = 0 \tag{21}$$

where $\boldsymbol{v}_A(\boldsymbol{r},t)$ is the flux or the net flow of polymer out of a small volume at position $\boldsymbol{r}$ and time t. Equation 21 is exact, but in order to make progress a relation between flux and concentration is required, which results in a closed expression. The simplest such relation is known as Fick's law in which the flux is proportional to the spatial gradient of the concentration, the physical reasoning of which is that at equilibrium the concentration is constant. However this is only true for non-interacting systems, which would not phase separate anyway. When interactions are present, it is not the concentration but the chemical potential of each component which is constant at equilibrium; hence a generalised version of Fick's law may be written,

$$J_A(r,t) = -\frac{M}{k_B T}\nabla\mu(r,t) \tag{22}$$

where M is a mobility and μ is the chemical potential difference defined by,

$$\mu(r,t) = \mu_A(r,t) - \mu_B(r,t) \tag{23}$$

The chemical potential difference is related to the free energy. Using the Flory Huggins theory modified to include energetic costs for gradients in concentration, the free energy is given by,

$$\begin{aligned} F &= \int \mathrm{d}r\big(f_{\mathrm{mix}}(r,t) + f_{\mathrm{gradient}}(r,t)\big) \\ &= \int \mathrm{d}r \frac{\phi_A}{N_A}\ln\phi_A + \frac{\phi_B}{N_B}\ln\phi_B + \chi\phi_A\phi_B + \kappa\left(\nabla\phi_A\right)^2 \end{aligned} \tag{24}$$

where $\chi = a + b/T$ is the temperature dependent Flory - Huggins interaction parameter and κ is the interfacial tension. Since F is a functional integral, $\nabla\mu$ in Eq. 22 must be replaced by a functional derivative, which physically corresponds to the total change in the free energy due to a small change in composition at r',

$$\begin{aligned} \mu(r',t) &= \frac{\delta F}{\delta\phi_A(r')} \equiv \lim_{\varepsilon\to 0}\frac{F\left[\phi_A + \varepsilon\delta\left(r - r'\right)\right] - F\left[\phi_A\right]}{\varepsilon} \\ &= \frac{\partial f_{\mathrm{mix}}(r',t)}{\partial\phi_A(r',t)} - 2\kappa\nabla^2\phi_A(r',t) \end{aligned} \tag{25}$$

where $\delta\left(r - r'\right)$ is the Kroneker delta function. Substitution of Eq. 25 into Eq. 22 results in a non-linear partial differential equation that cannot be solved analytically without further approximation. There are two approaches to making progress; one is to solve the equation of motion numerically, often using the finite difference method, the other is to linearise with respect to ϕ_A. The former approach is particularly valuable for exploring structural development, but the latter as we will see below can be used to determine stability limits and predict scattering patterns, both of which can be accessed experimentally. Linearisation involves expanding the concentration as a Taylor series, and retaining only terms proportional to $\delta\phi_A$, so that the chemical potential can be expanded as,

$$\mu(r,t) \approx \text{constant} + 2\left[\left(\chi - \chi_S\right) - \kappa\nabla^2\right]\delta\phi_A(r,t) + \cdots \tag{26}$$

where χ_S is the value of the interaction parameter on the quiescent spinodal,

$$\chi_S = \frac{1}{2}\left(\frac{1}{N_A\phi_A} + \frac{1}{N_B\phi_B}\right) \tag{27}$$

Assuming that the mobility is not spatially dependent, the resulting linearised equation of motion,

$$\frac{\partial \phi_A(r,t)}{\partial t} - 2M\nabla^2\left[\left(\chi - \chi_S\right) - \kappa\nabla^2\right]\delta\phi_A(r,t) = 0 \tag{28}$$

can be cast in a form that is straightforward for analytical manipulation by use of the Fourier transform,

$$f(q) = \frac{1}{(2\pi)^{3/2}} \int_V \mathrm{d}r f(r) \exp\{-\,\mathrm{i}q \cdot r\} \tag{29}$$

The result is a first-order linear differential equation,

$$\frac{\partial \delta\phi_A\,(q,t)}{\partial t} =-\, 2q^2 M\left[\left(\chi_S - \chi\right) + \kappa q^2\right]\delta\phi_A(q,t) \tag{30}$$

We have effectively decomposed the real space fluctuations into an uncoupled spectrum of fluctuations, each with wavevector, q. Equation 30 has the simple solution,

$$\delta\phi_A\left(q,t\right) = \delta\phi_A\left(q,t_0\right)\exp\left\{-\,2q^2 M\left[\left(\chi_S - \chi\right) + \kappa q^2\right]t\right\} \tag{31}$$

Hence spontaneous random fluctuations decay exponentially with a well-defined wavevector dependence. Equation 31 can also be used to help us understand the early stages of phase separation, since following a jump into the two-phase region, $\chi_S < \chi$, and for small enough values of q the decay rate becomes negative and fluctuations will therefore grow exponentially. By using the dependence of the interfacial tension derived by deGennes [41], $\kappa = b^2/36\phi(-\,1\phi)$, the well-known predictions of the Cahn–Hilliard theory for a binary monodisperse blend undergoing phase separation, follow. Firstly, there is a fastest growing wavevector, $q_m^2 = (\chi - \chi_S)/2\kappa$, which from Eq. 32 can be seen to grow at a rate $R(q_m) = M(\chi_S - \chi)^2/2$. It is this which gives rise to the characteristic co-continuous morphology associated with spinodal decomposition. Secondly, for wavevectors above a critical value, $q_c = \sqrt{2}q_m$, the rate becomes negative and such fluctuations decay. Concentration fluctuations can be probed by scattering techniques, which measure the structure factor,

$$S(q,t) = \left\langle|\delta\phi_A(q,t)|^2\right\rangle \tag{32}$$

In the one-phase region of the phase diagram, the Cahn–Hilliard theory predicts that $S(q,t) \to 0$ as $t \to \infty$, which is incorrect since fluctuations are continually appearing and decaying due to thermal motion, so that the structure factor has a finite value at all q. To account for this Cook [40] added a random noise term, $\theta(r,t)$, which, in order to ensure the correct equilibrium behaviour, has the following properties,

$$\left\langle\theta\left(r,t\right)\right\rangle = 0\,; \quad \left\langle\theta\left(r,t\right)\theta\left(r',t'\right)\right\rangle = 2M\nabla^2\delta\left(r - r'\right)\delta\left(t - t'\right) \tag{33}$$

or, after Fourier transformation,

$$\langle \theta(q,t) \rangle = 0; \quad \langle \theta(q,t)\,\theta(q,t') \rangle = 2Mq^2\delta(t-t') \tag{34}$$

The addition of a noise term to Eq. 30 means that a solution is only possible if it is converted into an equation of motion for the structure factor, using the definition of Eq. 32, and the average properties of the noise term, Eq. 34, are applied. The result is an evolution equation for the structure factor,

$$\frac{\partial S(q,t)}{\partial t} = -4q^2M\left[(\chi_S - \chi) + \kappa q^2\right]S(q,t) + 2Mq^2 \tag{35}$$

which in the steady state, $\partial S(q,t)/\partial t = 0$, predicts the correct form of the static structure factor, $S^{-1}(q) = 2\left(\chi_S - \chi + \kappa q^2\right)$. This latter expression can be derived using the random phase approximation (RPA) [17], which has proven extremely successful for predicting equilibrium scattering from polymer mixtures in the one-phase region. The RPA relates the collective structure factor to the structure factor of individual non-interacting chains and the Flory-Huggins interaction parameter, and is based on the idea that interactions act as a perturbation to ideal behaviour.

The simplest extension of the Cahn–Hilliard–Cook model to include shear is to assume that the only consequence is that concentration fluctuations are convected with the macroscopic flow field, as suggested by Lai and Fuller [43], adopting a model proposed by Ohta et al. [44]. With such an assumption, stress, and hence flow rate, gradients are neglected. This is only likely to be a physically realistic assumption if the viscoelastic properties of the two components are identical. The equation of motion is then a simple modification of Eq. 28,

$$\begin{aligned}&\frac{\partial \delta\phi_A(r,t)}{\partial t} - \dot{\gamma} y \hat{x} \nabla \delta\phi_A(r,t) \\ &= 2M\nabla^2\left[(\chi - \chi_S) - \kappa\nabla^2\right]\delta\phi_A(r,t) + \theta(r,t)\end{aligned} \tag{36}$$

which, after applying the Fourier transform, becomes,

$$\begin{aligned}&\frac{\partial \delta\phi_A(q,t)}{\partial t} - \dot{\gamma} q_x \frac{\partial \delta\phi_A(q,t)}{\partial q_{x,y}} \\ &= -2Mq^2\left(\chi_S - \chi + \kappa q^2\right)\delta\phi_A(q,t) + \theta(q,t)\end{aligned} \tag{37}$$

The structure factor then evolves according to,

$$\begin{aligned}&\frac{\partial S(q,t)}{\partial t} - \dot{\gamma} q_x \frac{\partial S(q,t)}{\partial q_y} \\ &= -4q^2M\left[(\chi_S - \chi) + \kappa q^2\right]S(q,t) + 2Mq^2\end{aligned} \tag{38}$$

The second term on the left hand side of Eqs. 36 and 38 represents the convective effect of shear. The steady state solution to Eq. 38, is,

$$2\left[q^2\left(\chi_S - \chi + \kappa q^2\right) - \dot{\gamma} q_x \frac{\partial}{\partial q_y}\right] S(q,t) = q^2 \tag{39}$$

Since this is still a differential equation, albeit in the variable q_y, a simple interpretation is not straightforward. Hence it is necessary to consider the time dependence of Eq. 38, which can be found using the method of characteristics. Through a transformation of variables, a first-order partial differential equation is converted into a first-order ordinary differential equation, which can then be solved using standard techniques. Due to the usefulness of this method for solving equations describing coupled shear flow and concentration fluctuation dynamics, it is worth briefly outlining the ideas. If we introduce a variable, t', such that,

$$\frac{\mathrm{d}t}{\mathrm{d}t'} = 1\,; \quad \frac{\mathrm{d}q_y}{\mathrm{d}t'} = -\dot{\gamma} q_x \tag{40}$$

and make use of the relation,

$$\frac{\mathrm{d}S}{\mathrm{d}t'} = \frac{\mathrm{d}t}{\mathrm{d}t'}\frac{\partial S}{\partial t} + \frac{\mathrm{d}q_y}{\mathrm{d}t'}\frac{\partial S}{\partial q_y} \tag{41}$$

then Eq. 38 can be rewritten as,

$$\frac{\mathrm{d}S}{\mathrm{d}t} = -4q^2(t)M\left[\left(\chi_S - \chi\right) + \kappa q^2(t)\right]S + 2Mq^2(t) \tag{42}$$

where we have dropped the $'$ notation for convenience, and $\boldsymbol{q} = (q_x, q_y, q_z)$ evolves according to the solution of Eq. 40,

$$\boldsymbol{q}(t) = \boldsymbol{q}_0 - \dot{\gamma} t q_{0x} \hat{\boldsymbol{q}}_y \tag{43}$$

The subscript 0 denotes the values of the wavevector at $t = 0$. The method of characteristics is not only valuable from a mathematical viewpoint; it also allows insight into the physical implications of the model. For example, an important consequence of Eq. 43 is that that any fluctuation with a nonzero q_x component, will eventually be convected to large values of $\boldsymbol{q}$. As we have already discussed, during the early stages of phase separation, only fluctuations with a wavevector less than a critical value grow. Hence, eventually all fluctuations with $q_x \neq 0$ will be convected to a regime in which they decay. Furthermore, it is clear that if a fluctuation appears with $q_{0x} = 0$, then the right-hand side of Eq. 42 becomes independent of time, and the structure factor evolves in exactly the same manner as if no shear were applied.

It is worth noting that several years prior to the approach adopted Lai and Fuller [43], Pistoor and Binder [45] had already attempted to include molecular effects of shear within the Cahn–Hilliard theory. They utilised the random phase approximation which offers an alternative formulism for determining the expansion of the chemical potential, in terms of local deviations from the mean concentration, which as noted in the previous section is an essential ingredient of the Cahn–Hilliard theory. In reference [45], the structure factor of individual non-interacting chains under the influence of a shear flow was first calculated assuming that the chains deformed affinely, i.e., with the flow. As a consequence the chemical potential becomes explicitly, rather than only implicitly due to convection, dependent on the shear rate. Their equation of motion for the structure factor took a modified form of Eq. 42,

$$\frac{dS}{dt} = 2Mq^2(t) - 4q^2(t)M\left[(\chi_S - \chi) + \kappa\left\{q^2(t) + \frac{2\pi^2}{15}\dot{\gamma}\tau q_x q_y(t) + \frac{4\pi^4}{315}\left(\dot{\gamma}\tau q_x\right)^2\right\}\right]S \tag{44}$$

where τ is a relaxation time related to the relaxation times and compositions of each of the components. Since only non-interacting chains were considered the model is only applicable to un-entangled chains that obey Rouse dynamics. A further consequence of this assumption is that since the chain stretching is irrespective of the environment there is no additional driving force for phase separation, and Eq. 44 only results in modified predictions for the structure factor rather than stability.

From the above discussion, it is clear that convection with or without an environment independent response of chains to shear is not enough to shift the conditions for stability of concentration fluctuations in the one-phase region. Its usefulness as a theory lies in its ability to predict the anisotropic q dependence of the structure factor in the one-phase region, which can be measured using scattering techniques. In order to predict apparent shifts in phase boundaries an extra ingredient is required – stress and flow rate variations within the blend.

The principal works in which the effects of stress gradients on the dynamics of such fluctuations in polymer solutions have been considered are by Helfand and Fredrickson [46], Onuki [47, 48], and Milner [49]. In all references the physical ideas are very similar, with variations only in the details of the theory developed.

Doi and Onuki [50] (DO), extended the models to polymer blends in which both components are entangled. The key aspect to address is how to incorporate stress into the equation of motion for concentration fluctuations. Effectively, by determining the conditions for force balance, it was shown that the stress enters the equation of motion at the same level as the chemical potential. Such an approach enabled the development of a framework that coupled the dynamics of concentration fluctuations to the flow fields and stress gradients; however, only the simplest form of constitutive relation for the stress was treated. In entangled polymer solutions, the tube model predicts that the relaxation of an imposed stress is well described by a single exponential decay, with the characteristic time-scale being that required for a polymer to reptate entirely out of its original tube [21]. Hence DO assumed that the dynamics of stress relaxation were well described by a single relaxation time.

A significant feature of the rheological behaviour of polymer blends is that, in general, stress relaxation cannot be adequately described using a single relaxation time. Essentially simultaneous publications by Sun et al. [51] (SBJ) and Clarke and McLeish. [52] (CM) attempted to address this shortcoming. SBJ proposed a model in which the shear modulus of the blend was allowed to be concentration dependent, whilst CM proposed the use of a specific mixing rule, which also included finite relaxation times for each of the components. In both models stress was incorporated into the modified Cahn–Hilliard equation of motion for concentration fluctuations derived by DO. The CM model has the advantage of being able to quantifiably predict that the effect of shear on the behaviour of the stability of concentration fluctuations in polymer blends. These changes lead to shifts in the spinodal curve, as defined dynamically rather than thermodynamically, of polymer blends. Dynamically the point at which small fluctuations first become unstable and grow defines the spinodal. This of course coincides with the thermodynamic definition in the quiescent state, but in the presence of shear enables us to consider apparent shifts in the spinodal without invoking unjustified thermodynamic concepts. Significantly, the dynamic equations also enable fluctuations in the flow field to be incorporated and determined. Since the flow itself is anisotropic it is to be expected that variations in the flow field will also vary with direction. As we will show below, it is simplest to understand, from the viewpoint of stability, the behaviour in the shear gradient and vorticity directions. In steady state shear the only stresses that act in these directions are the normal forces; hence, the existence of normal forces is the key to explaining experimental observations. This is perhaps the major similarity between the dynamic and the thermodynamic approach; however, the dynamic approach results in a greater degree of subtlety. More precisely, it is gradients in the normal stress, due to gradients in the concentration, which alter the conditions for the growth of small concentration fluctuations from that of the quiescent blend. A second important difference is that the dynamic approach considers enhancement and stability of concentration fluctuations rather than coexistence of phases in a de-mixed blend. In this respect it can be argued that the thermodynamic approach has greater potential for predicting experimentally useful and technologically essential parameters; however there remain a number of questions with regards to the fundamental physical basis of the concept.

In polymer solutions, both the longest relaxation time associated with the polymer and the plateau modulus scale with concentration, although the exact scaling law has yet to be established. The consequent increase in the first normal stress with concentration leads to the prediction that fluctuations in the shear gradient direction become unstable at temperatures within the quiescent one phase region, whereas in the vorticity direction, fluctuations remain stable for a range of temperatures within the quiescent two-phase region. In entangled polymer blends, the normal stress is dependent on the relative concentration of each component, with a relationship that is more complex than for solutions.

3.2.2
The Doi-Onuki Model

In order to understand the essential physical phenomena that occur when blends are sheared, DO utilized Rayleighs variational principle, which amounts to minimizing the energy dissipation function with respect to the average velocities of the two components. There are three contributions to the energy dissipation: friction between polymers, rate of change of the free energy due to concentration fluctuations and rate of change of the free energy due to stress. The friction was determined using the two-fluid model of Brochard [53]. In the tube model each polymer, with a degree of polymerisation, N, moves through a fixed network with a center of mass velocity, v, which can be related to the curvilinear velocity of a chain, w, within its tube. When a polymer segment has diffused a distance L, the contour length of the tube, the center of mass of the chain has only diffused a distance R, the end-to-end distance of the chain. Hence,

$$v = \frac{R}{L} w \tag{45}$$

The tube comprises N/N_e entanglement segments, where N_e is the average degree of polymerisation between entanglements, such that the mean-squared end-to-end distance is given by,

$$\langle R^2 \rangle = \frac{N}{N_e} \langle L^2 \rangle \tag{46}$$

Hence the center of mass velocity of a chain through the network is,

$$v^2 = \frac{N_e}{N} w^2 \tag{47}$$

To account for the fact that the polymers are not moving in a fixed network, Brochard introduced the concept of the tube velocity, v_T, which is the velocity of a hypothetical moving network, and is common to all polymers. Consequently, the chains now move with a velocity given by,

$$v = \frac{R}{L} w + v_T \tag{48}$$

The energy dissipated during this process is proportional to the square of the *curvilinear* velocity, and can be written as,

$$W_{\text{friction}} = \int \mathrm{d}r \left[\phi_A \zeta_{0A} w_A^2 + \phi_B \zeta_{0B} w_B^2\right] \tag{49}$$
$$= \int \mathrm{d}r \left[\phi_A \frac{N_A}{N_{eA}} \zeta_{0A} \left(v_A - v_T\right)^2 + \phi_B \frac{N_B}{N_{eB}} \zeta_{0B} \left(v_B - v_T\right)^2\right]$$

where ζ_{0i} is the monomeric friction coefficient. For simplicity it has, to date, always been assumed that is equal for each component; i.e., $\zeta_{0A} = \zeta_{0A} = \zeta_0$. In the second line, the energy dissipation is written in terms of the average velocities rather than the curvilinear velocities, since these are the natural variables when considering diffusion processes. It is clearly necessary to self-consistently eliminate the molecular field, v_T. DO proposed that this can done by minimizing Eq. 49 with respect to the tube velocity, $\delta W/\delta v_T = 0$, which amounts to balancing the forces on the network. Hence,

$$v_T = \frac{\zeta_A v_A + \zeta_B v_B}{\zeta_A + \zeta_B} \tag{50}$$

whereζ_i is the frictional drag of each polymer, given by,

$$\zeta_i = \phi_i \frac{N_i}{N_{ei}} \zeta_0 \tag{51}$$

The appearance and decay of concentration fluctuations results in energy dissipation, which can be determined by the rate of change of the corresponding free energy fluctuations,

$$W_{\text{mix}} = \dot{F} = \int \mathrm{d}r \frac{\partial F}{\partial \phi_A} \dot{\phi}_A(r,t) \equiv \int \mathrm{d}r \mu_A \dot{\phi}_A(r,t) \tag{52}$$

The last equality serves to define μ_A, a quantity that is similar, but not identical to the chemical potential. After substitution of the continuity equation into Eq. 52, it is found that,

$$W_{\text{mix}} = - \int \mathrm{d}r \mu_A \nabla \cdot \left(v_A \phi_A\right) \tag{53}$$

The stress dissipation is incorporated by determining the resultant change in the free energy when a small instantaneous deformation $r \to r + u(r)$ is imposed,

$$W_{\text{el}} = \frac{\partial \delta F_{\text{el}}}{\partial t} = \frac{\partial}{\partial t} \int \mathrm{d}r \sigma_{\alpha\beta} \frac{\partial u_\alpha}{\partial r_\beta} = \int \mathrm{d}r \sigma_{\alpha\beta} \frac{\partial v_{T\alpha}}{\partial r_\beta} \equiv \int \mathrm{d}r \sigma : \nabla v_T \tag{54}$$

This formalism is similar to the thermodynamic approach, and is based on the same principle that strain in polymeric systems results in a change in the free energy, which can be determined by calculating the work done. An important difference is that all components of the stress, and not just the first normal stress, are incorporated. DO also proposed that the stress arises from the motion of the tube, and thus is coupled to the tube velocity rather than the macroscopic flow velocity. An interesting consequence of this assumption is that stresses arise even when no macroscopic flow is imposed. This so called viscoelastic effect has been explored in some detail and a comprehensive review by Tanaka can be found in reference [54]. An closely related implication is an effective isotropic stress at non zero q. This results in unusual effects, such as a q dependent kinetic coefficient in the early stages of spinodal decomposition [55, 56]. When determining stability limits, it is only effects that occur in the limit of small q that are important. Hence, it is only necessary to consider the effects of the macroscopic velocity.

Equations 49, 53 and 54 are combined to give the total dissipation function. The minimization needs to be carried out with the additional constraint of incompressibility,

$$\nabla \cdot v = \nabla \cdot (\phi_A v_A + \phi_B v_B) = 0 \tag{55}$$

Hence, the function to be minimized with respect to each of the average polymer velocities is,

$$W_{\text{friction}} + W_{\text{mix}} + W_{\text{el}} + \int dr p(r,t) \nabla \cdot (\phi_A v_A + \phi_B v_B) \tag{56}$$

where $p(r,t)$ is a Lagrange multiplier. The minimization results in expressions for the total forces acting on component A,

$$\frac{\zeta_A \zeta_B}{\zeta_A + \zeta_B}(v_A - v_B) + \phi_A \nabla \mu_A + \phi_A \nabla p - \frac{\zeta_A}{\zeta_A + \zeta_B} \nabla \cdot \sigma^{(n)} = 0 \tag{57}$$

and those acting on component B,

$$\frac{\zeta_A \zeta_B}{\zeta_B + \zeta_A}(v_B - v_A) + \phi_B \nabla p - \frac{\zeta_B}{\zeta_A + \zeta_B} \nabla \cdot \sigma^{(n)} = 0 \tag{58}$$

The nomenclature $\sigma^{(n)}$ has been introduced to clarify that it represents the total network stress. By eliminating the Lagrange multiplier from Eqs. 57 and 58, and inserting the resultant expression into the continuity equation, DO arrived at an equation of motion for concentration fluctuations that included the effects of stress,

$$\frac{\partial \phi_A(r,t)}{\partial t} = - \nabla \cdot (\phi_A^2 v_A + \phi_A \phi_B v_B) + \nabla (\phi_A^2 \phi_B^2 / \zeta) \left[\nabla \mu_A - \alpha \nabla \cdot \sigma^{(n)} \right] \tag{59}$$

where,

$$\zeta = \frac{\zeta_A \zeta_B}{\zeta_B + \zeta_A} \tag{60}$$

and,

$$\alpha = \frac{(N_A/N_{eA})\zeta_0^A - (N_B/N_{eB})\zeta_0^B}{(N_A/N_{eA})\phi_A \zeta_0^A + (N_B/N_{eB})(1 - \phi_A)\zeta_0^B} \tag{61}$$

The importance of this prefactor will be discussed later.

An important simplification in terms of enabling analytical progress is to assume that the stresses are given by expressions appropriate to the macroscopically imposed flow field, and that the stress at any point, $\boldsymbol{r}$, in the system depends only on the local concentration. That is, when the concentration changes, it is assumed that the stress instantaneously changes; the dynamic evolution of the stress is assumed to be much faster than that for the concentration fluctuations. This is the so-called adiabatic approximation [49].

Linearisation and Fourier transformation of Eq. 59 results in the required equation of motion, for a fluctuation, in q space, $\delta\phi_{\mathrm{A}}(q,t)$,

$$\frac{\partial\delta\phi_{\mathrm{A}}(\boldsymbol{q},t)}{\partial t} - \dot{\gamma} q_x \frac{\partial\delta\phi_{\mathrm{A}}(\boldsymbol{q},t)}{\partial q_y} = - M\left[2q^2(\chi_{\mathrm{S}} - \chi + \kappa q^2) - \frac{\alpha}{k_{\mathrm{B}}T}\sum_{i,j} q_i q_j \frac{\partial}{\partial\phi_{\mathrm{A}}}\sigma_{ij}^{(n)}\right]\delta\phi_{\mathrm{A}}(\boldsymbol{q},t) \tag{62}$$

Without stress gradients, i.e., for a blend whose components have identical responses to shear, Eq. 62 reduces to Eq. 36, as would be expected. The natural progression from the model of Lai and Fuller [43] is evident if Eq. 62 is compared with Eq. 37.

3.2.3
Stability Analysis

In order to determine the conditions for fluctuations to grow or decay, at the level of approximations discussed above, it is only necessary to consider changes in the condition for the growth or decay of fluctuations in the $q_x = 0$ plane, and it is sufficient to focus our attention on the q_y and q_z directions. In these directions, the convective term of Eq. 62 vanishes, and an effective diffusion coefficient, D_{eff} may be defined using,

$$\frac{\partial\delta\phi_{\mathrm{A}}(q_{y,z},t)}{\partial t} =- q_{y,z}^2 D_{\mathrm{eff}}\delta\phi_{\mathrm{A}}(q_{y,z},t) \tag{63}$$

which is no more than the Cahn–Hilliard [39] equation for the dynamics of concentration fluctuations with a modified diffusion coefficient which depends on gradients in the stress as well as gradients in the chemical potential. If thermal noise is neglected, Eq. 63 has the simple solution,

$$\delta\phi_{\mathrm{A}}(q_{y,z},t) = \delta\phi_{\mathrm{A}}(q_{y,z},t=0)\exp\left\{- q_{y,z}^2 D_{\mathrm{eff}} t\right\} \tag{64}$$

An apparent spinodal can be defined by the condition $D_{\mathrm{eff}} = 0$ for $q \to 0$. If $D_{\mathrm{eff}} < 0$, fluctuations become unstable and grow, rather than decay.

As a consequence of the approximations made, Eq. 63 is a linearised equation in $\delta\phi_{\mathrm{A}}(q,t)$; hence, the model is not able to predict the behaviour as second and higher-order fluctuation effects become important once the fluctuations have grown beyond a certain magnitude.

3.2.4 The Asymmetric Shear Response of Polymer Blends

A key assumption of DO is that, although there are two components, there is only one network stress. The contribution of each component to this stress should however be determined from some mixing rule which relates blend dynamics to that of the pure components.

An alternative approach to including the complexity of blend rheology was considered by CM. At low enough shear rates; i.e., $\dot{\gamma}\tau < 1$ (where τ is the longest molecular relaxation time in the blend), a general description of the relation between stress and strain is provided by the upper convective Maxwell model, Eq. 5. From the viewpoint of constructing a model for polymer blends, an important and useful consequence of this model, is that the overall stress factors into a term due to the molecular dynamics and a term due to the flow field. In reference [52], it was proposed that the molecular response of the blend, as determined by the stress relaxation function, $G(t - t')$, can be most simply modeled using the double reptation concept introduced by des Cloiseaux [57] and independently by Tsenoglou [58]. Although each approached the problem from slightly different physical viewpoints, the consequences are similar. A discussion of the differences between the two approaches and the good agreement between the model and experiment can be found in reference [59]. A word of caution is necessary at this point; the model has been developed and verified for blends of chemically identical polymers with different molecular weights, whereas in references [52] and [60], it was assumed that it is also applicable to blends of chemically dissimilar polymers. This would only be true if the monomeric friction coefficients for each component do not depend on concentration.

Physically, double reptation accounts, in a simple way, for the fact that polymers do not reptate in fixed tubes, as assumed in the original tube model. The surrounding polymers, which form the tube, also relax, and so the constraints which form the tube decay with a characteristic time-scale. In other words, stress relaxation depends not only the dynamics of each individual polymer, but also on the dynamics of the surrounding polymers. It should be noted that more sophisticated and therefore complex rheological models for polymer mixtures have been proposed; however, a particular appeal of this model is the simple relation between the stress relaxation function of the blend and the Doi-Edwards stress relaxation function for polymers in a fixed network,

$$G(t) = \left[\phi_A \{G_A F_A(t)\}^{1/2} + \phi_B \{G_B F_B(t)\}^{1/2}\right]^2 \tag{65}$$

where G_i is the pure component plateau modulus of component i. $F_i(t)$ is the relaxation function, which, in the case of a homogeneous blend, physically corresponds to the probability of tube survival for component i in a matrix of identical relaxing obstacles. As was shown by Doi and Edwards [21], the probability of tube survival in a matrix of fixed obstacles is almost single exponential; hence, the dynamics are dominated by a single timescale, the terminal relaxation time. Consequently, for the double reptation model we have [59],

$$G(t) = \left[\phi_A\{G_A \exp(-t/\tau_A)\}^{1/2} + \phi_B\{G_B \exp(-t/\tau_B)\}^{1/2}\right]^2 \tag{66}$$

τ_i represents some relaxation time, of component i, which in terms of the tube model, is related to the idealised Doi-Edwards relaxation time for component i in a matrix of fixed obstacles, τ_{DE}, by, $\tau_i = (1/2)\tau_{DE}$. Hence, in the double reptation model, the effect of constraint release is to half the relaxation time (if single exponential decay is assumed), from that predicted for a polymer in a fixed matrix. In the heterogeneous blends considered here, the τ_i are the tube survival times for chains of species i in an idealised environment, in which the chemical heterogeneity matches that of the blend, but all chains share the same relaxation time. That is, double reptation accounts for mutual effects in topological stress relaxation, but not for direct effects of local composition on the monomeric friction factors. The parameters of the double reptation model should be treated as phenomenological, to be determined from independent linear rheology experiments in the one phase region (see for example reference [61]).

In the steady state, Eqs. 11 and 14 clearly lead to expressions for the shear stress, σ_{xy}, and the first normal stress, N_1, which are functions of concentration,

$$\begin{aligned}\sigma_{xy} &= \eta\dot{\gamma} \\ &= \dot{\gamma}\left[\phi_A^2 G_A \tau_A + 4\phi_A(1-\phi_A)(G_A G_B)^{1/2}\left(\frac{\tau_A \tau_B}{\tau_A + \tau_B}\right) + \phi_B^2 G_B \tau_B\right]\end{aligned} \tag{67}$$

and,

$$\begin{aligned}N_1 &= \psi_1\dot{\gamma}^2 \\ &= 2\dot{\gamma}^2\left[\phi_A^2 G_A \tau_A^2 + 8\phi_A(1-\phi_A)(G_A G_B)^{1/2}\left(\frac{\tau_A \tau_B}{\tau_A + \tau_B}\right)^2 + {\phi_B}^2 G_B {\tau_B}^2\right]\end{aligned} \tag{68}$$

The Maxwell model results in a second normal stress, $N_2 = \sigma_{yy} - \sigma_{zz}$, equal to zero. In determining the contribution of the stress field to the effective diffusion coefficient, we also need to consider how the shear rate is perturbed by the presence of concentration fluctuations. If the relative velocities are eliminated from Eqs. 57 and 58, then an important constraint on the stress tensor can be found,

$$\nabla \times \nabla\sigma^{(n)} = 0 \tag{69}$$

Consequently, in the velocity gradient direction the shear stress is constrained to be constant, so that,

$$\delta\sigma_{xy}(q_y) = \left.\frac{\partial\sigma_{xy}}{\partial\phi_A}\right|_{\dot{\gamma}} \delta\phi_A + \left.\frac{\partial\sigma_{xy}}{\partial\dot{\gamma}}\right|_{\phi_A} \delta\phi_A = 0 \tag{70}$$

Hence, the variation of the shear rate in the velocity gradient direction with respect to the concentration is,

$$\left.\frac{\partial\dot{\gamma}}{\partial\phi_A}\right|_{\sigma_{xy}} =- \left.\frac{\partial\sigma_{xy}}{\partial\phi_A}\right|_{\dot{\gamma}} \Big/ \left.\frac{\partial\sigma_{xy}}{\partial\dot{\gamma}}\right|_{\phi_A} \tag{71}$$

as was first proposed by Onuki [48] for polymer solutions. By contrast, in the vorticity direction the shear stress may vary, and no perturbation of the shear field is required to satisfy Eq. 18. It is these differences between the behaviour in two directions that gives an indication of the importance of using a dynamic approach.

From Eqs. 62 and 63, the effective diffusion coefficients in the q_y and q_z directions are given by,

$$\begin{aligned} D_{\text{eff}}(q_i) &= 2M(\chi_c - \chi + \kappa q_i^2) - \frac{M}{k_B T}\frac{\partial\sigma_{ii}}{\partial\phi_A} \\ &\equiv 2M\left[\chi_c - \chi + \kappa q_i^2 + \Delta\chi_c(\vec{q}_i)\right] \quad (i = y, z) \end{aligned} \tag{72}$$

which serves to define $\Delta\chi_c(q_y)$ and $\Delta\chi_c(q_z)$. If $\Delta\chi_c > 0$ fluctuations are suppressed, whereas if $\Delta\chi_c < 0$ fluctuations are enhanced. In the latter case, the magnitude of $\Delta\chi_c$ may be sufficient for D_{eff} to become negative so that fluctuations become unstable and grow.

Hence, the effective diffusion coefficient depends on the choice of σ_{yy} and σ_{zz}. Rheological constitutive equations are only enable the stress tensor to be determined to within an isotropic constant. DO argued briefly that, in the tube model, if the length of a polymer is equal to its equilibrium value, so that stress arises from orientation, $\text{Tr}\,\sigma = 0$. For example, if the diagonal components of the solution to the constitutive equation for the stress tensor are given by $\sigma'_{xx} = N_1, \sigma'_{yy} = N_2, \sigma'_{zz} = 0$, then the appropriate components of the stress tensor should be written, $\sigma'_{xx} = N_1 + c, \sigma'_{yy} = N_2 + c, \sigma'_{zz} = c$, so that,

$$\sigma_{xx} = \frac{2}{3}N_1 - \frac{1}{3}N_2\,; \quad \sigma_{yy} = \frac{2}{3}N_2 - \frac{1}{3}N_1\,; \quad \sigma_{zz} =- \frac{1}{3}N_1 \tag{73}$$

By comparison, HF, considered the enhancement of fluctuations in polymer solutions using a constitutive equation of the form,

$$\boldsymbol{\sigma} = 2\eta\boldsymbol{D} - \psi_1\boldsymbol{D}^{\nabla} + 4\psi_2\boldsymbol{D}^2 \tag{74}$$

where,

$$\boldsymbol{D} = \frac{1}{2}\left(\nabla\boldsymbol{v} + \nabla\boldsymbol{v}^{\text{T}}\right) \tag{75}$$

and,

$$D^{\nabla} \equiv \frac{\partial}{\partial t}\sigma + \boldsymbol{v}\cdot(\nabla \boldsymbol{D}) - (\nabla \boldsymbol{v})^{\mathrm{T}}\cdot \boldsymbol{D} - \boldsymbol{D}\cdot(\nabla \boldsymbol{v}) \tag{76}$$

ψ_2 is the second normal stress coefficient, which is often taken to be zero for polymer solutions and blends. Equation 74 is known as a *second-order fluid* [62], and it leads to,

$$\sigma_{xy} = \eta\dot{\gamma},\ \sigma_{xx} = \psi_1\dot{\gamma}^2,\ \sigma_{yy} = \psi_2\dot{\gamma}^2,\ \sigma_{zz} = 0 \tag{77}$$

In considering the consequences of such stresses on sheared polymer solutions HF did not assume that $\mathrm{Tr}\,\sigma = 0$. Clearly, whether this condition is imposed will have significant consequences for the predicted effects of shear flow on blends, as noted in reference [38]. In reference [60], it was assumed that the argument put forward by DO was applicable and hence determined the shift in the stability conditions was determined using Eq. 73, which leads to,

$$\begin{aligned}\Delta\chi_{\mathrm{c}}(q_y) = {} & \frac{2\alpha}{3k_{\mathrm{B}}T}\dot{\gamma}^2 G_{\mathrm{A}}\tau_{\mathrm{A}}^2 \\ & \times\Bigg[\left\{\phi_{\mathrm{A}} + 4(1-2\phi_{\mathrm{A}})\sqrt{G'}\left(\frac{\tau'}{1+\tau'}\right)^2 - (1-\phi_{\mathrm{A}})G'\tau'^2\right\} \\ & - \left[\phi_{\mathrm{A}}^2\tau_{\mathrm{A}}^2 + 8\phi_{\mathrm{A}}(1-\phi_{\mathrm{A}})\sqrt{G'}\left(\frac{\tau'}{1+\tau'}\right)^2 + \phi_{\mathrm{B}}^2 G'\tau_{\mathrm{B}}^2\right] \\ & \times\frac{\phi_{\mathrm{A}} + 2(1-2\phi_{\mathrm{A}})\sqrt{G'}\left(\frac{\tau'}{1+\tau'}\right) - (1-\phi_{\mathrm{A}})G'\tau'}{\phi_{\mathrm{A}}^2\tau_{\mathrm{A}} + 4\phi_{\mathrm{A}}(1-\phi_{\mathrm{A}})\sqrt{G'}\left(\frac{\tau'}{1+\tau'}\right) + \phi_{\mathrm{B}}^2 G'\tau'}\Bigg]\end{aligned} \tag{78}$$

and

$$\begin{aligned}\Delta\chi_{\mathrm{c}}(q_z) = {} & \frac{2\alpha}{3k_{\mathrm{B}}T}\dot{\gamma}^2 G_{\mathrm{A}}\tau_{\mathrm{A}}^2 \\ & \times\left\{\phi_{\mathrm{A}} + 4(1-2\phi_{\mathrm{A}})\sqrt{G'}\left(\frac{\tau'}{1+\tau'}\right)^2 - (1-\phi_{\mathrm{A}})G'\tau'\right\}\end{aligned} \tag{79}$$

These expressions show that the important dimensionless parameters in the problem are the relative modulae, $G' = G_{\mathrm{B}}/G_{\mathrm{A}}$, and the relative relaxation times, $\tau' = \tau_{\mathrm{B}}/\tau_{\mathrm{A}}$. If, on the other hand, Eq. 77 is used then there is no shift in the stability condition in the vorticity direction; i.e., $\Delta\chi_{\mathrm{c}}(q_z) = 0$ whilst in the flow-gradient direction,

$$\Delta\chi_{\mathrm{c}}(q_y) = \frac{2\alpha}{k_{\mathrm{B}}T}\left[\dot{\gamma}^2\frac{\partial\psi_2}{\partial\phi_{\mathrm{A}}} + 2\psi_2\frac{\partial\dot{\gamma}}{\partial\phi_{\mathrm{A}}}\right] \tag{80}$$

which is also zero for the weakly non-linear Maxwell model, since $\psi_2 = 0$. That experimental observations show that the stability condition does depend on shear rate in the vorticity direction provides some support for the argument of DO leading to the use of Eq. 73.

In reference [51], SBJ explore a phenomenological version of the DO theory. However, their starting point is less microscopic than DO. Firstly, rather than proposing the existence of a self-consistent field, such as the tube velocity, and hence determining the friction using the ideas that lead to Eq. 45 to Eq. 49, they assume, that the dissipation due to friction is given by,

$$W_{\text{friction}} = \int \mathrm{d}r \phi_A \zeta_0 \left(v_A - v_B\right)^2 \tag{81}$$

A second crucial difference, which again arises from the lack of a tube velocity, is that they assume that there is a stress associated with each component, rather than a network stress which is common to both, which couples to the individual velocities rather than the tube velocity. The resultant contribution to the total dissipation due to deformation can be written (cf. Eq. 54) as,

$$W_{\text{el}} = \int \mathrm{d}r \left[\sigma_A : \nabla v_A + \sigma_B : \nabla v_B\right] \tag{82}$$

It was assumed that the individual stresses are related to the total stress by,

$$\sigma_i = \phi_i \frac{G_i\left(\phi_i\right)}{G} \sigma \tag{83}$$

where $G_i(\phi_i)/G$ is a normalized modulus, for which no explicit form was given, of component i. Equations 81, 82 and 83 can then be substituted into Eq. 56 and a similar procedure as outlined above results in a slightly modified form of the shear dependent diffusion equation. When then considering the consequences of their model, they did not impose the condition that $\mathrm{Tr}\,\sigma = 0$, and also assumed the existence of an isotropic stress. This latter was derived on the basis of rubber elasticity theory rather than viscoelasticity and hence its applicability to polymer blends is questionable.

Criado-Sancho et al. [38] (CSJCV) extended the DO and CM formulism, but also assumed that $\mathrm{Tr}\,\sigma \neq 0$, arguing that such an assumption is valid for Couette type flows, if not cone and plate type flows. In addition they included an isotropic stress as a direct contribution to the free energy, and hence the chemical potential, in an analogous way to the theory of Wolf [23] discussed earlier, but within the framework of EIT. Their excess free energy took the form,

$$E_s = \Psi_1 \dot{\gamma}^2 = (x_A V_A + x_B V_B) J_e^0 \sigma_{xy}^2 \tag{84}$$

which is similar to Eq. 11, but does not include the non-linear shear effects. The double reptation mixing rule, rather than the more phenomenological mixing rule of Eq. 17, was used to determine the shear stress Eq. 67 and the steady state compliance,

$$J_e^0 = \frac{\phi_A^2 G_A \tau_A^2 + 8\phi_A(1-\phi_A)(G_A G_B)^{1/2}\left[\tau_A\tau_B/(\tau_A+\tau_B)\right]^2 + \phi_B^2 G_B \tau_B^2}{\phi_A^2 G_A \tau_A + 4\phi_A(1-\phi_A)(G_A G_B)^{1/2}\left[\tau_A\tau_B/(\tau_A+\tau_B)\right] + \phi_B^2 G_B \tau_B} \tag{85}$$

This latter expression follows from Eq. 14 relating the first normal stress and the shear stress. Unlike SBJ, directional dependence of the shear-induced spinodal shifts was incorporated by invoking the consequences of Eq. 69; i.e., in the flow-gradient direction the constraint of a constant shear stress was imposed, whereas in the vorticity direction the constraint of constant shear rate was imposed. This leads to expressions for $\Delta\chi_c(q_y)$ and $\Delta\chi_c(q_z)$ that are even more complex than those from the CM model, and so will not be reproduced here. However, a comparison of the predictions of the two different approaches was made as can be seen in Fig. 4. In the flow gradient direction CSJCV predict a more complex range of shear-induced stability changes than that predicted by the CM model.

In reference [52] the important role that the prefactor α plays in determining the apparent phase behaviour was neglected, which results in an inconsistency in the theory. From Eqs. 78 and 79, it is clear that whether $\Delta\chi_c$ is negative or positive depends also on the sign of α. For blends in which $N_{eA} = N_{eB}$, α is positive for $N_A > N_B$, and negative otherwise. This ensures symmetry; in other words the effects of shear are not dependent on the way in which the components are labelled. However, for $N_{eA} \neq N_{eB}$, the situation is modified; α is positive if,

$$\frac{N_A}{N_{eA}} > \frac{N_B}{N_{eB}} \tag{86}$$

It is useful to rewrite this condition in terms of the component relaxation times and plateau moduli. If the component relaxation times are given by the tube model, then,

$$\tau' = \frac{N_B^3 N_{eA}}{N_A^3 N_{eB}} \tag{87}$$

and the plateau modulus per monomer volume is given by,

$$\frac{G_i}{k_B T} \approx \frac{b_i^2}{a_i^2} = \frac{1}{N_{ei}} \tag{88}$$

where b is the statistical segment length and a is the tube diameter, and we have used the relation $a^2 \approx N_e b^2$. In other words, the energy arising from shear is $k_B T$ per entanglement segment, and the energy per monomer volume is $k_B T$ divided by the number of statistical segments per entanglement segment. An important point to recognize is that the use of Eq. 88 ensures consistency between the magnitudes of the Flory- Huggins and the shear stress contributions to the thermodynamic driving force for the dynamics of

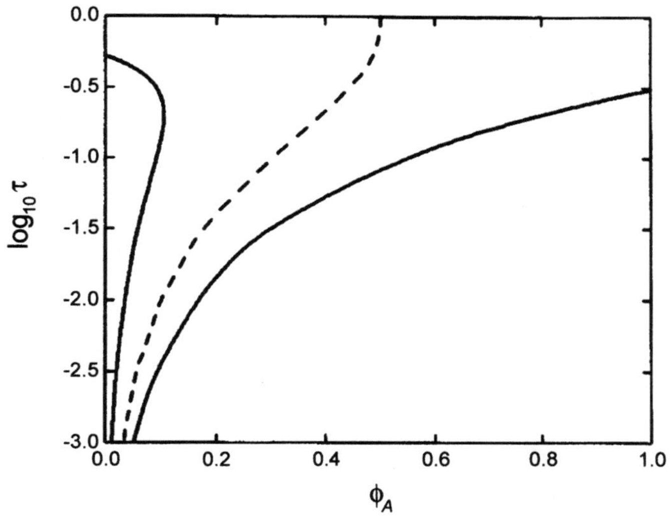

Fig. 4 Borders of instability in the flow-gradient direction for $G' = 1$, according to the model of Clarke and McLeish (*dotted line*), and Criado-Sancho et al. (*solid line*). The CM model predicts shear-induced suppression of fluctuations, i.e., $\Delta\chi_c > 0$, to the left of the dotted line and shear-induced enhancement to the right, i.e., $\Delta\chi_c < 0$. The CSJVC model predicts shear-induced suppression of fluctuations to the left of the left-most solid curve and to the right of the right-most solid curve and shear-induced enhancement between the two curves. Reproduced with permission from reference [38]

concentration fluctuations. Hence, the sign of α is determined by,

$$\alpha < 0 \text{ if } \tau' > 1/G'^4; \; \alpha > 0 \text{ if } \tau' < 1/G'^4 \tag{89}$$

Combining equations the relevant results enables predictions of the conditions for which $\Delta\chi_c(q_y) < 0$ and $\Delta\chi_c(q_y) > 0$ as a function of τ', concentration and G'. Figure 4 (dotted line) and Fig. 5 illustrate the behaviour in the flow-gradient direction for $G' = 1$ and $G' = 2$, respectively. In each case there are regions in which shear-induced mixing, $\Delta\chi_c(q_y) > 0$, can occur and regions in which shear-induced de-mixing, $\Delta\chi_c(q_y) > 0$, can occur. Although the behaviour is slightly less varied in the vorticity direction, there are again regions in which shear-induced mixing can occur and regions in which shear-induced de-mixing, can occur. In particular we noted that in order to observe shear-induced de-mixing, a necessary (but not sufficient) condition is that, if the components are labelled such that G' is greater than one, we must have τ' less than one. Of course, it is not simply the sign of $\Delta\chi_c$ which determines whether concentration fluctuations will grow or decay, it is also its magnitude with respect to the quiescent driving force, which is determined by the quantity, $\chi_c - \chi$.

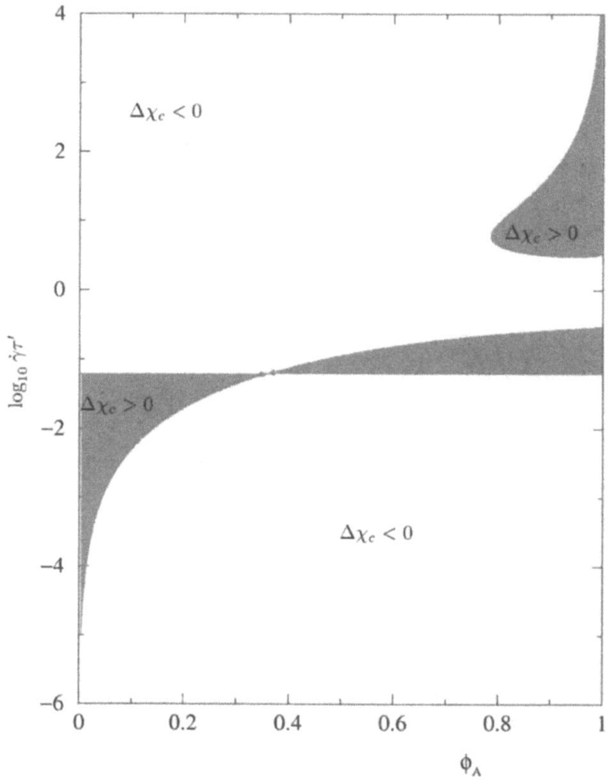

Fig. 5 Regions of shear-induced enhancement of concentration fluctuations, $\Delta\chi_c < 0$, for $G' = 2$, according to the CM model. Reproduced with permission from reference [60]

3.2.5 Non-linear Flow

Since for most entangled polymeric systems, the relaxation times are such that the condition $\dot{\gamma}\tau > 1$ is easily attained experimentally, it is natural to ask what will happen at higher shear rates, for which the Maxwell constitutive equation no longer applies. The only extension to date that includes such effects within the dynamic framework was introduced by CM and considered in detail by Clarke [60]. The theory reviewed in the previous section predicts that it is only the magnitude of the shift which changes with a change in shear rate; the qualitative effect remains the same, either shear-induced mixing or shear-induced de-mixing, depending on the particular set of values of τ', G' and ϕ_A. It is important to note that although the theory has been developed assuming small concentration fluctuations, it is not restricted to low shear rates only.

Doi and Edwards derived a constitutive equation for polymer melts, by applying the tube model for the regime of strong flow. They assumed that after deformation a polymer recovers its equilibrium length quickly by retracting within its tube; whereas the stress associated with the orientation of the tube with the flow remains after this initial retraction, and is relaxed more slowly by reptation. The stress relaxation function of Eq. 4, is the same as before; it represents the time dependent correlation function of the end to end vector of a reptating polymer chain. In other words, it is also possible to apply the mixing rule of des Cloiseaux to the strong flow regime, from which,

$$\sigma_{xy}^{\text{tot}} = \phi_{\text{A}}^2 G_{\text{A}} \sigma_{xy}^{\text{AA}} + 2\phi_{\text{A}}\phi_{\text{B}}\sqrt{G_{\text{A}}G_{\text{B}}}\sigma_{xy}^{\text{AB}} + \phi_{\text{B}}^2 G_{\text{B}} \sigma_{xy}^{\text{BB}} \tag{90}$$

and,

$$N_1^{\text{tot}} = \phi_{\text{A}}^2 G_{\text{A}} N_1^{\text{AA}} + 2\phi_{\text{A}}\phi_{\text{B}}\sqrt{G_{\text{A}}G_{\text{B}}}N_1^{\text{AB}} + \phi_{\text{B}}^2 G_{\text{B}} N_1^{\text{BB}} \tag{91}$$

In order to determine the three dimensionless shear stresses and dimensionless first normal stress differences, which are functions of the shear rate and relaxation times, that contribute to Eqs. 90 and 91 we utilised a model due to Larson [63], who showed that a differential approximation to Eq. 5 is given by,

$$\tau\left[\frac{\partial}{\partial t}\sigma + 2\boldsymbol{D}:\sigma\sigma - \nabla\boldsymbol{v}^{\text{T}}\cdot\sigma - \sigma\cdot\nabla\boldsymbol{v}\right] + \sigma = \delta \tag{92}$$

At low shear rates; i.e., $\dot{\gamma}\tau < 1$, the second term on the left hand side of Eq. 92 can be neglected, and the expression becomes an exact differential version of the upper convected Maxwell model. In steady state shear flow Eq. 92 gives each component first normal stress as,

$$N_1^{ii} = \frac{2\dot{\gamma}\tau\sigma_{xy}^{ii}}{1 + 2\dot{\gamma}\tau\sigma_{xy}^{ii}/3} \tag{93}$$

and the individual shear stresses as the solution to the cubic equation,

$$\frac{4}{9}(\dot{\gamma}\tau^{ii})^2\left(\sigma_{xy}^{ii}\right)^3 + \frac{2}{3}\dot{\gamma}\tau^{ii}\left(\sigma_{xy}^{ii}\right)^2 + \left(1 + \frac{2}{3}\dot{\gamma}\tau^{ii}\right)\sigma_{xy}^{ii} = \dot{\gamma}\tau^{ii} \tag{94}$$

where $\tau^{\text{AA}} \equiv \tau_{\text{A}}$, $\tau^{\text{BB}} \equiv \tau_{\text{B}}$ and $\tau^{\text{AB}} \equiv 2\tau_{\text{A}}\tau_{\text{B}}/(\tau_{\text{A}} + \tau_{\text{B}})$. The resultant shear stresses and first normal stress can then be substituted into Eq. 62 to give effective diffusion coefficients in the q_y and q_z directions. In the q_y direction the quantity $\partial\sigma_{xy}/\partial\dot{\gamma}$ is also required; this is easily found from Eq. 94.

A significant problem with the Doi-Edwards formulation (and Larson's approximation) is that a shear stress which decreases with shear rate, above some critical value of $\dot{\gamma}\tau$, is predicted. This is physically unrealistic, and is believed to be due to the model assuming that in strong flow the tube segments become too highly oriented in the direction of flow. Models which attempt to overcome this problem have only very recently been proposed [64, 65]. Consequently, it is necessary to restrict modelling in the q_y direction to shear rates for which shear thinning does not occur; i.e.,

$$\frac{\partial \sigma_{xy}(\max[\tau_A, \tau_B])}{\partial \dot{\gamma}} > 0 \tag{95}$$

The model also predicts a first normal stress which is constant in the limit of very strong flows. Again this represents too strong a degree of shear thinning. However, the model does allow us to examine, at least qualitatively, the effects of shear flow, in the shear thinning regime, on the behaviour of concentration fluctuations. The effect in the q_y direction for $G' = 1$ was illustrated in Fig. 2 of reference [52], whilst in reference [60] we presented results for the experimentally more accessible q_z direction. In the limit of very strong shear, for which the first normal stress difference becomes a constant, dependent only on the plateau modulus,

$$\lim_{\dot{\gamma}\to\infty} N_1^{ii} = 3G^{ii} \tag{96}$$

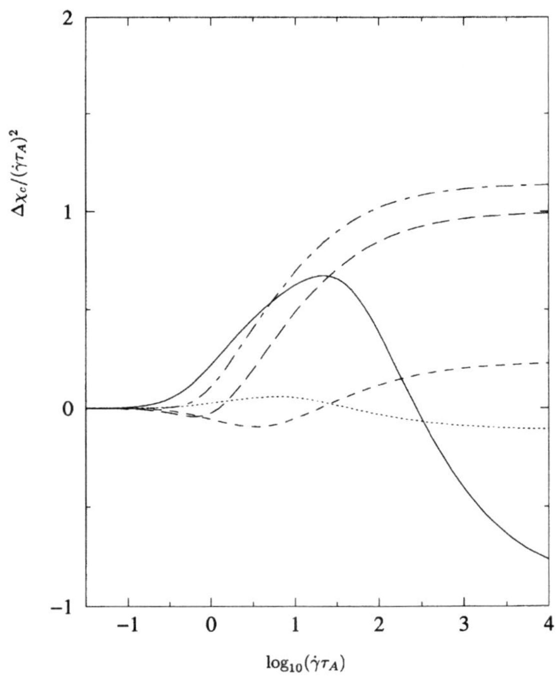

Fig. 6 The magnitude of $\Delta\chi_c$ as a function of shear rate for $G' = 2$ and $\phi_A = 0.5$, for various values of τ': (– · – · –) $\tau' = 1.0$, (— — —) $\tau' = 0.5$, (- - -) $\tau' = 0.1$, (······) $\tau' = 0.05$, and (—) $\tau' = 0.01$. As can be seen for $\tau' = 1.0$, $\Delta\chi_c$ is always positive; for $\tau' = 0.5$ and $\tau' = 0.1$, $\Delta\chi_c$ changes from being negative at low shear rates to positive at higher shear rates; and for $\tau' = 0.05$, and $\tau' = 0.01$, $\Delta\chi_c$ changes from positive at low shear rates to negative at higher shear rates. Reproduced from permission from reference [60]

where $G^{AA} \equiv G_A$, $G^{BB} \equiv G_B$ and $G^{AB} \equiv \sqrt{G_A G_B}$, the shift in the spinodal is simply given by,

$$\Delta\chi_c(q_z) = \frac{2\alpha G_A}{k_B T}\left(\phi_A + \phi_B\sqrt{G'}\right)\left(1 - \sqrt{G'}\right) \quad (97)$$

As a consequence, whether shear-induced mixing or de-mixing occurs in the strong shear limit depends only on G' and α. For $G' = 1$, $\Delta\chi_c(q_z)$, which is always positive at low shear rates, becomes zero in the strong shear limit. For $G' \neq 1$, the result of Eq. 97 can be combined with the predictions of the previous section. A range of complex shear dependent behaviours were predicted.

In Fig. 6, we illustrate the magnitude of the shear-induced shift as a function of shear rate for $G' = 2$ and $\phi_A = 0.5$. Various values of τ' are chosen to illustrate the range of possible behaviours, including shear-induced mixing at low shear rates but shear-induced de-mixing at higher shear rates, and visa versa. In each case it can be seen that the magnitude increases dramatically above $\dot{\gamma}\tau_A \approx 0.1$, showing that a consideration of non-linear flow effects is important.

3.2.6 Model Systems

It is desirable to find a model system which overcomes the problem of some of the assumptions made in the theory. Probably the most serious problem arises from the assumption that the monomeric friction coefficients are equal for both components and that they do not vary as a function of composition. The most appealing candidates for which this assumption can be expected to be reasonable are blends of chemically identical polymers, in which one component has been deuterated. As shown by Wiltzius et al. [66] for blends of polybutadiene (PB) and deuterated PB, and Beaucage et al. [67] for blends of polydimethylsiloxane (PDMS) and deuterated PDMS, such polymers still possess an interaction parameter and can undergo phase separation. The advantage of both these candidates is that they possesses a phase separation temperature above their respective glass transition temperatures, for high enough molecular weights. Both Clarke [60] and more recently CSJCV have applied their respective theories to such model blends.

First, we note that in order to predict a phase diagram as a function of shear rate, we must account for the variation with temperature of the molecular dynamics. It is well established that the terminal relaxation time changes rapidly with temperature, due to changes in the monomeric friction coefficient, and a suitable description of the behaviour is provided by the phenomenological WLF formula [68],

$$\zeta_0(T) = \zeta_0(T_0)a_T \quad (98)$$

where a_T is the shift factor,

$$\log a_T = \frac{-c_1^0(T - T_0)}{c_2^0 + T - T_0} \tag{99}$$

T_0 is some reference temperature and c_1^0 and c_2^0 are empirical constants. As noted previously, the terminal relaxation time is related to the Doi-Edwards relaxation time of a polymer in a fixed matrix; hence [21],

$$\tau_i = \frac{5\zeta_0 b^2 N_i^3}{8k_B T\pi^2 N_e} \tag{100}$$

and τ' is given by Eq. 87. The quiescent spinodal curve is determined by Eq. 27, and for a PB/dPB blend the relation between χ and temperature was found to be [66],

$$\chi \approx \frac{0.198}{T} \tag{101}$$

The statistical segment length of PB is [69] $b = 6.9$ angstroms; the degree of polymerisation of an entanglement segment is [68] $N_e = 35$; the monomeric friction coefficient at a reference temperature, $T_0 = 298$ K, is [68] $\zeta_0 = 1.607 \times 10^{-10}$ dynes-sec/m; the WLF coefficients, at the same reference temperature are [68] $c_1^0 = 3.64$ and $c_2^0 = 186.5$ K.

The resultant phase diagrams for a particular molecular weight combination are shown in Fig. 7 for the q_z direction. The most striking observation was that the phase boundary in the q_z direction is much more sensitive to shear than that in the q_y direction. This is because the variation of the shear rate which occurs in the shear gradient direction reduces, to a certain degree, the gradient in the stress.

The effect of there being a maximum in the shear stress as $\dot{\gamma}$ increases, was shown in reference [60], where at higher shear rates, and lower temperatures (where the monomeric friction coefficient is greatest), the effective spinodal curve diverged away from the quiescent spinodal. From Eq. 71, it is clear that such a maximum would lead to a singularity in the effective diffusion coefficient. Since a maximum in the shear stress is not in line with experimental observations [62], we should also be aware that the prediction of the spinodal diverging from the quiescent curve is an artefact of the constitutive equation used to model the behaviour of polymers in strong flows.

From Fig. 7, we see that it is possible to produce "islands" of immiscibility, and even for the region of immiscibility to disappear completely (within the illustrated shear range) at intermediate shear rates. Since $G' = 1$, we can see from Eq. 97 that in the limit of strong flow, the model predicts that the effect of shear will vanish. This occurs because in this limit there are no stress gradients; the stress is independent of the degree of polymerisation, and is dependent only on the plateau modulus.

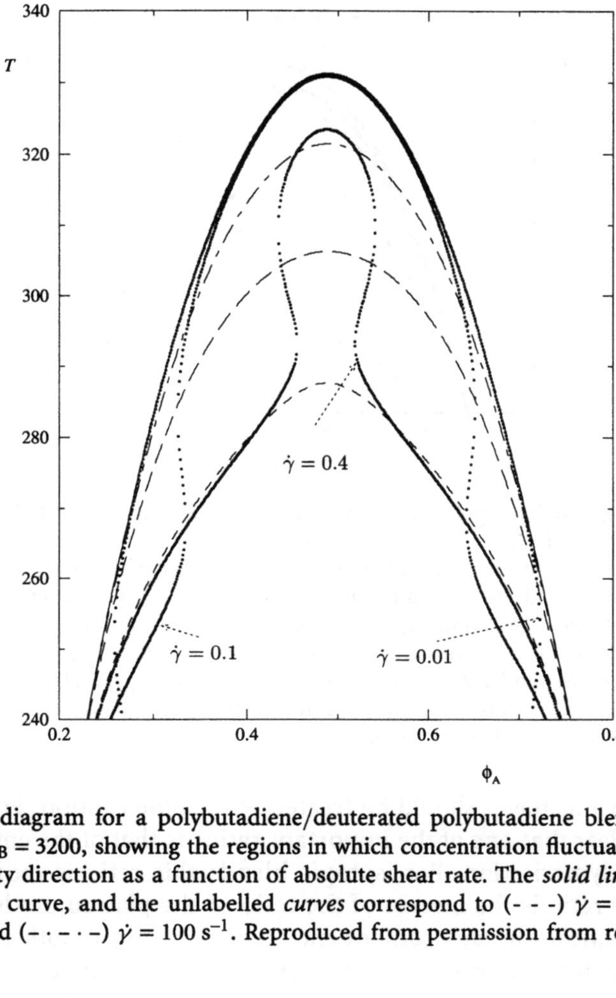

Fig. 7 Phase diagram for a polybutadiene/deuterated polybutadiene blend with $N_{PB} = 3500$ and $N_{dPB} = 3200$, showing the regions in which concentration fluctuations will grow in the vorticity direction as a function of absolute shear rate. The *solid line* is the quiescent spinodal curve, and the unlabelled *curves* correspond to (- - -) $\dot{\gamma} = 0.5\ s^{-1}$, (– – –) $\dot{\gamma} = 10\ s^{-1}$, and (– · – · –) $\dot{\gamma} = 100\ s^{-1}$. Reproduced from permission from reference [60]

In reference [38] it was assumed that the relaxation time was independent of temperature over a range of 100 °C which is unrealistic. A second difference between the predictions is that Clarke allowed non-linear rheological behaviour, which becomes especially important as the temperature is decreased due to a corresponding increase in the relaxation time, whereas CSJCV only considered weakly non-linear rheological behaviour. Nonetheless the predictions of reference [38] are shown in Fig. 8 for a blend of PDMS/dPDMS in which the degree of polymerisation of each component is 964 and 957, respectively. Significantly the predictions are compared with those due to the CM model, and it is found that whereas the latter predicts shear-induced mixing in the vorticity direction, the CSJCV model leads to shear-induced de-mixing. Such differences highlight a potential method for experimentally determining the most appropriate model. However, the predictions according to the

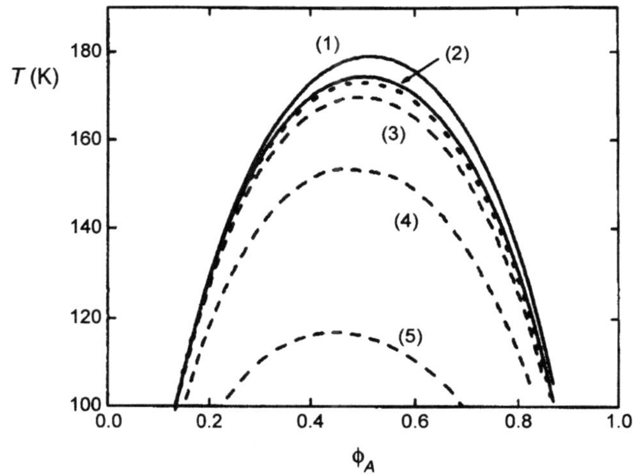

Fig. 8 Phase diagram for a polydimethylsiloxane (PDMS) and deuterated PDMS blend, with $N_{PDMS} = 964$ and $N_{dPDMS} = 957$, showing the regions in which concentration fluctuations will grow in the vorticity direction as a function of absolute shear rate. The *dotted curve* corresponds to the quiescent spinodal. The continuous curves are calculated according to the CSJCV model, and the *dashed curves* according to the CM model. Curves 1 and 5: $\dot{\gamma} = 10\,s^{-1}$, curves 2, 3 and 4: $\dot{\gamma} = 5\,s^{-1}$. Curve 3 was calculated by CSJVC using the CM model with a ratio of entanglement degree of polymerisation of $N_{ePDMS}/N_{edPDMS} = 0.9$, whereas for curves 4 and 5, $N_{ePDMS}/N_{edPDMS} = 0.5$ was used. Reproduced with permission from reference [38]

CM model for this blend should be treated with some caution. From Eqs. 87 and 88, it is clear that one of the important ratios is that of the entanglement degrees of polymerisation. For isotopic blends this ratio is most probably ~ 1, since the ratio is the inverse of the ratio of the plateau modulae, which should be identical for each component. In such a case if both components also have very similar degrees of polymerisation, the CM model predicts that shear will have no effect on stability. In order to overcome this problem in their comparison between the models, CSJVC calculated shifts according to the CM model with the ratio of the entanglement degrees of polymerisation $\neq 1$. This illustrates another difference between the models. The CM model relies on rheological asymmetry between components, whereas the CSJCV model predicts shear-induced phenomena even without such asymmetry.

3.2.7
Temperature Effects

In the above model calculations it was assumed that the various dimensionless rheological parameters are equal at all temperatures. Clarke and McLeish [70] extended their model to include different temperature dependencies of the rheological parameters. Although many of the characteristics of the blend vary with temperature, it was proposed that the most significant factor is the temperature dependence of the monomeric friction coefficient, ζ_0^i. In pure melts, ζ_0 changes rapidly with temperature, and a suitable description of the behaviour is given by Eqs. 98 and 99.

The situation in blends of different polymers is much less well understood, despite the attention that has recently been devoted to the subject [71–76]. It is generally observed that blends in the miscible state are thermorheologically complex; i.e., it is not possible to determine a single set of WLF parameters that will lead to time-temperature superposition at all frequencies. Colby [71] suggested that a reasonable description of the monomeric friction coefficient for each component in the blend can be obtained by using c_1^0 and c_2^0 as measured for the pure components, and T_g as measured for the blend. Consequently, the local dynamics of the two components are coupled through the use of a single T_g. However, each component friction factor has its own distinct temperature dependence, which is in agreement with observations [71, 77, 78]. More sophisticated theoretical schemes have been suggested [79, 80], which propose that local concentration fluctuations result in a range of local dynamic environments. Such an idea is successful in describing the broad range of glass transition temperatures that are observed in polymer blends. However, these models remain phenomenological, in that it is still assumed that each component retains its own WLF parameters upon blending [80]. We used the original idea proposed by Colby since it captures the underlying physical behaviour in a manner suitable for quantitative analysis.

The important consequence of applying Eq. 99, using pure component values for c_1^0 and c_2^0, and the blend T_g, is that the ratio of the monomeric friction coefficients is temperature dependent. This in turn leads to temperature dependence of both α (see Eq. 61), and τ'. If we assume the statistical segment lengths of each component to be equal, then the response of the blend to shear flow is characterised by three independent parameters for each component, the degree of polymerisation of an entanglement segment, N_{ie}, the degree of polymerisation of a chain, N_i, and the monomeric friction coefficient, ζ_0^i. Equivalently, we may use N_i, N_{ie}, and the terminal relaxation time τ_i, fixed at some arbitrary temperature, as our independent parameters. The temperature dependence of τ_i is then governed by the WLF equation, with the appropriate parameters.

We illustrated the possible consequences of such behaviour for a model system, based on the SCPE/EVA blend studied in reference [81], with $N_{SCPE} = 3300$ and $N_{EVA} = 1960$. This blend phase separates upon heating; that is it exhibits a lower critical solution temperature. We estimated the following relationship between χ and temperature, $\chi = 0.011 - (3.8/T)$, using experimentally determined spinodal temperatures for two different 50/50 SCPE/EVA blends [4]. The quiescent spinodal curve is found from the Flory-Huggins free energy. We also included a concentration dependence of the glass transition temperature, by use of the Fox-Flory relation [68],

$$\frac{1}{T_g} = \frac{\phi_A}{T_{gA}} + \frac{\phi_B}{T_{gB}} \tag{102}$$

where T_g is the glass transition temperature of the blend, and T_{gi} is the glass transition temperature of pure component i.

The outcome of substituting this temperature dependence into the effective diffusion coefficient in the vorticity direction given by equation is illustrated in Fig. 9, with the relevant model parameters given in reference [70]. For the stated WLF parameters τ' increases with temperature. For each curve, $\dot{\gamma}$ is constant for all temperatures and compositions, and the relaxation times of each component are fixed (arbitrarily) at $T = 355$ K. Since the magnitude of the shear dependent shift is proportional to $(\dot{\gamma}\tau_A)^2$, and since τ_A decreases rapidly with temperature the effect of shear is less noticeable at higher temperatures. This is illustrated by the smaller shifts in the phase boundary which occur at higher temperatures. The results clearly illustrate that for a small range of shear rates, it is possible to induce miscibility gaps, if the rela-

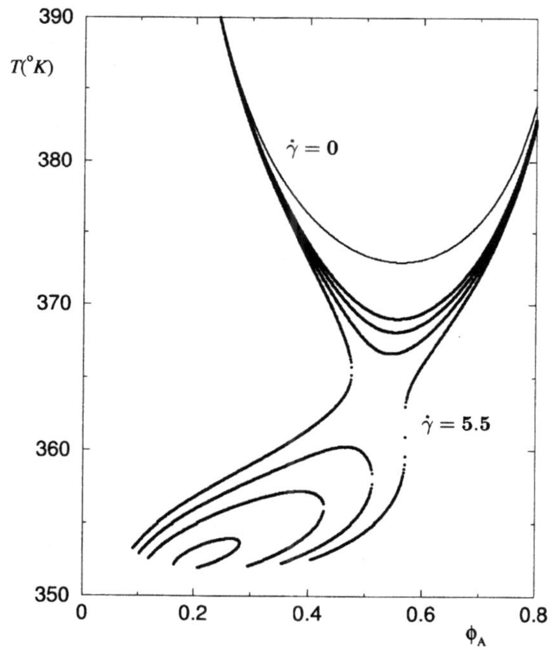

Fig. 9 The effect of shear on first-order fluctuations, for a blend of SCPE/EVA. The three unlabelled curves are $\dot{\gamma} = 4.75\ \text{s}^{-1}$ (for the inner most miscibility gap and the shifted spinodal closest to the $\dot{\gamma} = 0\ \text{s}^{-1}$ curve), $\dot{\gamma} = 5\ \text{s}^{-1}$ and $\dot{\gamma} = 5.25\ \text{s}^{-1}$. The behaviour for shear rates of $\dot{\gamma}\tau_A > 0.75\ \text{s}^{-1}$ was not calculated since at such reduced shear rates the weakly Newtonian theory we have used is not applicable. Reproduced with permission from reference [70]

tive relaxation time, τ', and the relative plateau modulus, G', satisfy certain criteria. As discussed earlier, a necessary, but not sufficient, condition being that if $G' > 1$, then $\tau' < 1$. A significant difference between our results and the experimental observations is the temperature range over which closed loop miscibility gaps occur. Experimentally [81], the range is only 10 degrees below the quiescent spinodal, our model suggests a typical range of 20–30 degrees. This difference may be due to a lack of accurate rheological data.

Whilst we accounted for a concentration dependence of the glass transition temperature, we did not include the additional contributions to the gradient of the stress which arise from gradients in the monomeric friction factor. Once we have determined the temperature dependence of the relaxation times using the appropriate blend T_g, we assume the relaxation times to be fixed and not a function of concentration. Hence, our theory is strictly only applicable to blends in which each component has a similar glass transition temperature. Despite attempts to understand and explain the consequences that such behaviour will have on blend rheology, a clear picture has yet to emerge.

Another possible factor that we neglected is the effect of shear flow on the glass transition temperature. Since the phase boundaries are well above the glass transition temperature, we may expect that this does not play too important a role.

Our results indicated that closed loop miscibility gaps are probably an exception, requiring a delicate balance of the governing parameters. In most blends, shear will simply shift the spinodal. The delicacy of the criteria necessary to explain closed loop miscibility gaps may go some way to explaining why only a few experimental systems have been found to display them. In relaxing the constraint of the ratio of the relaxation time being temperature independent, the theory represents an important step in developing a physical picture that accounts for the important underlying mechanisms driving the response of polymer blends to shear flow.

3.3 Scattering Under Shear

We have seen that scattering techniques are useful for probing shifts in phase boundaries; however they can provide more detailed information about organisation particularly as a phase boundary is approached. As already noted a phase separating blend often scatters light very strongly and becomes optically opaque. Even near a phase boundary the fluctuations, which provide the initial structures for growth once inside the two phase region, are very large. The theory presented in the previous section can be used to predict scattering patterns that are anisotropic, and thus measuring such patterns can test the theory further. The starting point is identical to that presented in the previous section: the equation of motion for the concentration fluctuations that includes convection and coupling to the stress, and a dynamic equation for the stress. Since we are now interested in scattering patterns we will no longer confine attention to the $q_y = 0$ and $q_z = 0$ directions.

As an illustration, we note that Eq. 38, can be solved through the method of the integrating factor,

$$S(q,t) = \exp\left\{\int_0^t D(q,t')\mathrm{d}t'\right\} \times \left[S_0(q) + \int_0^t C(q,t'')\exp\left\{-\int D(q,t')\mathrm{d}t'\right\}\mathrm{d}t''\right] \tag{103}$$

where the structure factor at time $t = 0$ is defined as,

$$S_0^{-1}(q) = 2\left(\chi_S - \chi + \kappa q_0^2\right) \tag{104}$$

Clearly numerical techniques are necessary to calculate the structure factor from Eq. 103, which is why to date such predictions (and indeed measurements) have been mostly confined to polymer solutions [49, 82]. An exception is due to Lai and Fuller [43], who by neglecting the thermal noise contribution were able to derive an analytical expression for the time dependence of the structure factor. As shown in Fig. 10, they showed that in the flow –

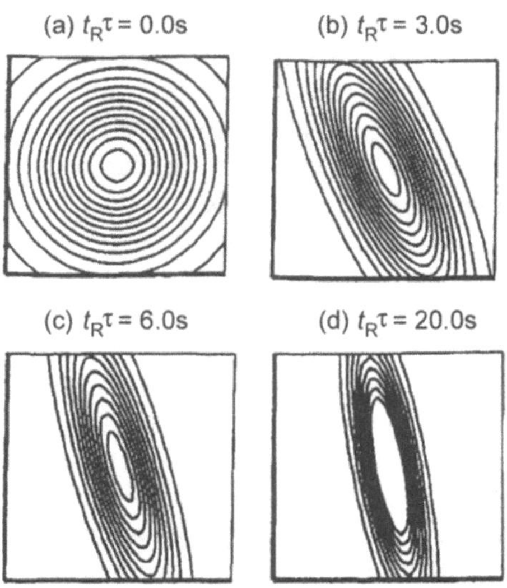

Fig. 10 Time resolved structure factor in the flow – flow gradient plane following the imposition of shear calculated according to the model of Lai and Fuller. $t_R\tau$ represents a scaled time

flow gradient plane the contours of the structure factor are ellipsoidal and oriented at an angle of 135° to the direction of flow immediately following the imposition of shear. The orientation angle then rotates with time towards the direction perpendicular to the flow. This was qualitatively in-line with what they observed experimentally, suggesting that a model which only includes the convective effects of shear is able to capture some important physical processes. However, as they noted, the mixture that they studied, polystyrene/polybutadiene/dioctylphthalate probably had a negligible viscoelastic nature.

The model of CM has recently been adapted by Miroshnychenko and Clarke [83] to predict enhanced scattering from entangled polymer solutions subjected to oscillatory flow, and a reasonable agreement between predictions and experiments [84] was found. This highlights the importance of scattering techniques in this field – they permit a quantitative comparison between theory and experiment.

3.4 Numerical Modelling

A natural progression from the linear stability analysis and the calculation of scattering curves is to use numerical techniques to solve the complete set of equations of motion including that for the evolution of the stress – an advantage of numerical modeling is that the adiabatic approximation is no longer necessary. Although a large number of simulations of binary mixtures under flow have been carried out (see for example references [85–96]), which have simultaneously solved diffusion and flow equations, very few have incorporated the effect that stress has on the diffusive flux, as discussed above. Hence the simulations are only able to predict the effects that shear has on morphology development, and the consequences of morphology on rheological response. Since it is clear from experiments that the coupling between stress and diffusion is vital to understand the behaviour of high molecular weight polymer blends, such works seem likely to have limited applicability to the viscous mixtures of interest in this review, and the results are restricted to systems where viscous effects are minimal or there is negligible viscosity difference between the components.

To date, only a limited number of studies incorporating the coupling of Eq. 59 have been carried out for either polymer solution or blends. Onuki et al. [97] used a time-dependent Ginzburg-Landau model to describe the two fluid dynamics of polymer and solvent, which is similar in physical content to Eq. 59. Stress evolution was determined using a form of Eq. 92 without the non-linear term (i.e., they used the upper convected Maxwell model). They found that above the coexistence curve; i.e., within the one phase region, a dynamical steady state was reached, with enhanced fluctuations. At relatively large shear elongated polymer-rich regions formed a transient network supporting most of the stress. An interesting, and experimentally verifiable, feature of their results was that because such a network is continuously deformed in shear flow, the shear stress and the normal stress difference exhibited large fluctuations. Such predictions are only possible using numerical techniques; the linear stability analysis assumes that the magnitude of fluctuations of shear and normal stresses are given once the concentration fluctuations are known. Okuzono [98] also numerically studied the dynamics of polymer solutions in two dimensions in the one-phase region under the influence of shear flow. Viscoelastic effects were incorporated somewhat differently into the model by introducing particles which had memories of their own past history. They found that shear-induced phase separation can change qualitative features of rheological response, which is unsurprising. Jupp et al. [99] have investigated the model introduced by Clarke and McLeish, in particular using the proposed relation, Eq. 66, between composition and stress. It is somewhat surprising given the extent to which their modelling relies on the theory of references [52] and [60], that they reference the former only in passing and the latter not at all. As well as modelling structural evolution and finding similar results to Onuki as regards the behaviour of the stresses, they numerically investigated the conditions for an instability. Since this can be done exactly using the analytical approach discussed in previous sections, this seems a somewhat cumbersome approach. A disadvantage of numerical techniques is that whilst they are often able to qualitatively reproduce observations, quantitative comparison is difficult since it is difficult to determine whether morphological differences are due to simply choosing the wrong parameters or the model being insufficiently physically realistic.

4 Conclusions

The effect of shear on entangled polymer blends is both qualitatively and quantitatively different from that on polymer solutions. Two different approaches to understanding the origins of such behaviour have emerged, one based purely on thermodynamic considerations, the other on dynamic considerations. That there is common ground between the models is indicated by the fact that within the framework of either, the controlling factors are the relaxation times and the plateau modulae. Importantly these are parameters that can be determined by independent rheological experiments. Recently the two approaches have begun to converge, as highlighted by the work of CSJCV. Although the various models have been able to qualitatively reproduce the varied experimental observations, much work is still necessary before a clear picture of the most suitable theoretical approach emerges. From this viewpoint, the most promising experimental observations are the shift in stability conditions and scattering patterns as instabilities are approached. Blends with components that have the most closely matched molecular parameters (e.g., the friction coefficient) and glass transition temperatures are those that are most likely to provide a strong test of the theoretical ideas.

References

1. Mazich KA, Carr SH (1983) J Appl Phys 54:5511
2. Mani S, Malone MF, Winter HH, Halary JL, Monnerie L (1991) Macromolecules 24:5451
3. Larbi FBC, Malone MF, Winter HH, Halary JL, Leviet MH, Monnerie L (1988) Macromolecules 21:3532
4. Hindawi IA, Higgins JS, Weiss RA (1992) Polymer 33:2522
5. Fernandez ML, Higgins JS, Horst R, Wolf BA (1995) Polymer 36:149
6. Hong Z, Shaw MT, Weiss RA (2000) Polymer 41:5895
7. Katsaros JD, Malone MF, Winter HH (1986) Polym Bull 16:83
8. Katsaros JD, Malone MF, Winter HH (1989) Polym Eng Sci 29:1434
9. Mani S, Malone MF, Winter HH (1992) Macromolecules 25:5671
10. Madbouly SA, Ougizawa T, Inoue T (1999) Macromolecules 32:5631
11. Hong ZY, Shaw MT, Weiss RA (1998) Macromolecules 31:6211
12. Gerard H, Higgins JS, Clarke N (1999) Macromolecules 32:5411
13. Gerard H, Cabral JT, Higgins JS (2003) Philos Trans R Soc Lond Ser A-Math Phys Eng Sci 361:767
14. Gerard H, Higgins JS (1999) PCCP Phys Chem Chem Phys 1:3059
15. Larson RG (1999) The Structure and Rheology of Complex Fluids. Oxford University Press, New York
16. de Gennes PG (1971) J Chem Phys 55:572
17. de Gennes PG (1980) Scaling Concepts in Polymer Physics. Ithaca, New York
18. Doi M, Edwards SF (1978) J Chem Soc Faraday Trans II 74:1789
19. Doi M, Edwards SF (1978) J Chem Soc Faraday Trans II 74:1802
20. Doi M, Edwards SF (1978) J Chem Soc Faraday Trans II 74:1818
21. Doi M, Edwards SF (1986) The Theory of Polymer Dynamics. Oxford University Press, Oxford
22. Horst R, Wolf BA (1992) Macromolecules 25:5291
23. Wolf BA (1984) Macromolecules 17:615
24. Graessley WW (1974) Adv Polym Sci 16:1
25. Schuch H (1988) Rheol Acta 27:384
26. Horst R, Wolf BA (1991) Macromolecules 24:2236
27. Horst R, Wolf BA (1993) Macromolecules 26:5676
28. Horst R, Wolf BA (1994) Rheol Acta 33:99
29. Krause C, Wolf BA (1997) Macromolecules 30:885
30. Krause C, Horst R, Wolf BA (1997) Macromolecules 30:890
31. Wolf BA (1999) J Chem Phys 110:7542
32. Horst R, Wolf BA (1997) Polymer 38:4697
33. An L, Hinrichs A, Horst R, Krause C, Wolf BA (2000) Macromol Symp 149:75
34. Sun ZY, Yang JA, Jiang W, An LJ, Jiang ZH, Wu ZW (2002) Polymer 43:4047

35. Chopra D, Haynes C, Hatzikiriakos SG, Vlassopoulos D (1999) J Non-Newton Fluid Mech 82:367
36. Jou D, Casasvazquez J, Criadosancho M (1995) Physical Properties of Polymers, p 207
37. Jou D, Casas-Vazquez J, Lebon G (1999) Rep Prog Phys 62:1035
38. Criado-Sancho M, Jou D, Casas-Vazquez J, del Castillo LF (2002) Phys Rev E 66:061803
39. Cahn JW, Hilliard JE (1958) J Chem Phys 28:258
40. Cook HE (1970) Acta Metall 18:297
41. de Gennes PG (1980) J Chem Phys 72:4756
42. Pincus P (1981) J Chem Phys 75:1996
43. Lai J, Fuller GG (1994) J Polymer Sci: Part B: Polymer Phys 32:2461
44. Ohta T, Nozaki H, Doi M (1990) J Chem Phys 93:2664
45. Pistoor N, Binder K (1989) Colloid Polym Sci 266:132
46. Helfand E, Fredrickson GH (1989) Phys Rev Lett 62:2468
47. Onuki A (1990) J Phys Soc Jpn 59:3423
48. Onuki A (1990) J Phys Soc Jpn 59:3427
49. Milner ST (1991) Phys Rev Lett 66:1477
50. Doi M, Onuki A (1992) J Phys II 2:1631
51. Sun T, Balazs AC, Jasnow D (1999) Phys Rev E 59:603
52. Clarke N, McLeish TCB (1998) Phys Rev E 57:R3731
53. Brochard F (1988) M Nagasawa (ed) Elsevier Science, Amsterdam p 249
54. Tanaka H (2000) J Physics-Condensed Matter 12:R207
55. Onuki A, Taniguchi T (1997) J Chem Phys 106:5761
56. Clarke N, McLeish TCB, Pavawongsak S, Higgins JS (1997) Macromolecules 30:4459
57. Descloizeaux J (1988) Europhys Lett 5:437
58. Tsenoglou C (1991) Macromolecules 24:1762
59. Mead DW (1996) J Rheol 40:633
60. Clarke N (1999) Faraday Discuss 112:249
61. Pavawongsak S, Higgins JS, Clarke N, McLeish TCB, Peiffer DG (2000) Polymer 41:757
62. Larson RG (1988) Constitutive Equations for Polymer Melts and Solutions. Butterworths, Boston
63. Larson RG (1984) J Rheol 28:545
64. Marucci G, Ianniruberto G (1997) In: Adams MJ, Mashelkar JR, Pearson JRA, Rennie AR (eds) Dynamics of complex fluids. Imperical College Press, The Royal Society, London
65. Larson RG, Mead DW, Doi M (1998) Macromolecules 31:7895
66. Wiltzius P, Bates FS, Heffner WR (1988) Phys Rev Lett 60:1538
67. Beaucage G, Sukumaran S, Clarson SJ, Kent MS, Schaefer DW (1996) Macromolecules 29:8349
68. Ferry JD (1988) Viscoelastic properties of polymers. Wiley, New York
69. Bates FS, Dierker SB, Wignall GD (1986) Macromolecules 19:1938
70. Clarke N, McLeish TCB (1999) Macromolecules 32:4447
71. Colby RH (1989) Polymer 30:1275
72. Zawada JA, Fuller GG, Colby RH, Fetters LJ, Roovers J (1994) Macromolecules 27:6851
73. Zawada JA, Fuller GG, Colby RH, Fetters LJ, Roovers J (1994) Macromolecules 27:6871
74. Kant R, Kumar SK, Colby RH (2003) Macromolecules 36:10087
75. Lodge TP, McLeish TCB (2000) Macromolecules 33:5278
76. Roovers J, Toporowski PM (1992) Macromolecules 25:1096
77. Chung GC, Kornfield JA, Smith SD (1994) Macromolecules 27:5729
78. Chung GC, Kornfield JA, Smith SD (1994) Macromolecules 27:964
79. Katana G, Fischer EW, Hack T, Abetz V, Kremer F (1995) Macromolecules 28:2714

80. Kumar SK, Colby RH, Anastasiadis SH, G Fytas (1996) J Chem Phys 105:3777
81. Fernandez ML, Higgins JS, Richardson SM (1996) J Mat Proc Tech 56:807
82. Ji H, Helfand E (1995) Macromolecules 28:3869
83. Miroshnychenko D, Clarke N (2002) Phys Rev E 66:041802
84. Saito S, Matsuzaka K, Hashimoto T (1999) Macromolecules 32:4879
85. Xu A, Gonnella G, Lamura A (2003) Phys Rev E 67:056105
86. Luo KF, Yang YL (2002) Macromol Theory Simul 11:429
87. Roths T, Friedrich C, Marth M, Honerkamp J (2002) Rheol Acta 41:211
88. Zhang ZL, Zhang HD, Yang YL (2001) J Chem Phys 115:7783
89. Berthier L (2001) Phys Rev E 63:051503
90. Lamura A, Gonnella G (2001) Physica A 294:295
91. Corberi F, Gonnella G, Lamura A (2000) Phys Rev E 62:8064
92. Zhang ZL, Zhang HD, Yang YL (2000) J Chem Phys 113:8348
93. Liu H, Chakrabarti A (2000) J Chem Phys 112:10582
94. Shou ZY, Chakrabarti A (2000) Phys Rev E 61:R2200
95. Wu YN, Skrdla H, Lookman T, Chen SY (1997) Physica A 239:428
96. Padilla P, Toxvaerd S (1997) J Chem Phys 106:2342
97. Onuki A, Yamamoto R, Taniguchi T (1997) J Phys II 7:295
98. Okuzono T (1997) Mod Phys Lett B 11:379
99. Jupp L, Kawakatsu T, Yuan XF (2003) J Chem Phys 119:6361

Author Index Volumes 101–183

Author Index Volumes 1–100 see Volume 100

Subject Index